I0070738

Why We Fall Apart

The Seven Cellular Sins Behind Ageing and Cancer

Dr. Sasi Shanmugam Senga

Why We Fall Apart
The Seven Cellular Sins of Ageing and Cancer

© **2026 Dr. Sasi Shanmugam Senga**

Published by **Kalavathi Cancer Foundation**

All rights reserved. No part of this publication may be reproduced, stored in a retrieval system, or transmitted in any form or by any means, electronic, mechanical, photocopying, recording, or otherwise, without the prior written permission of the author, except for brief quotations in reviews.

First published 2026

ISBN 978-0-6489285-1-5

A catalogue record for this book is available from the Library of Congress.

The author has asserted his right to be identified as the author of this work in accordance with the Copyright, Designs and Patents Act 1988.

This book is a work of scientific synthesis. It is not intended as medical advice and should not be used as a substitute for professional medical consultation.

CONTENTS

PART III - Redemption Without Illusion

Redemption as Rebalancing, Not Reversal
The Biological Meaning of Discipline
After the Sins

About the Author

Dr. Sasi Shanmugam Senga is a neurosurgical oncologist with a Master's degree in Neuroscience and a Master's degree in Cancer and Therapeutics. He is a UK Commonwealth Scholar, an Oxford Clarendon Scholar, and a recipient of the Harvard Excellence Award.

He has served as a lecturer in Medicine at the University of Oxford and the University of Buckingham, and as Programme Director in Molecular Genetics and Ethics at Stanford University. He has also contributed to international policy and research discussions as a panellist for *The Economist* think tanks.

In parallel with his academic work, he is involved in the evaluation of global higher-education systems and contributes annually to the QS World University Rankings for top universities worldwide.

He is an Ambassador and active member of leading international cancer organisations, including the European Association for Cancer Research, the European Society for Medical Oncology, the American Association for Cancer Research, and the American Society of Clinical Oncology.

Dr. Senga is the author of a Royal Society–top-cited research article, *Hallmarks of Cancer: The New Testament*. Alongside his academic and clinical work, He is also involved in the work of the Kalavathi Cancer Foundation, a charitable organisation supporting underprivileged women and children affected by cancer. His work lies at the intersection of cancer biology, ageing, neuroscience, ethics, and the limits of medical intervention.

In memory of Kalavathi

mother of the author,
whose life and death shaped the author's path toward
contributing, in however small a way, to the benefit of
humankind.

Why We Fall Apart: The Seven Cellular Sins Behind Ageing and Cancer

Introduction

Why We Fall Apart

Tempus edax rerum.
Time, the devourer of things.

We grow old even when nothing obvious goes wrong.

Not through catastrophe, nor through any single betrayal of the body, but by degrees so small they escape notice. Long before illness acquires a name, long before pain or diagnosis or decline, the body begins to lose a certain exactness. Repair slows. Signals blur. Cells that once behaved with exemplary discipline linger past their usefulness, divide when restraint would serve the whole, or refuse to die when departure would be kinder. Nothing dramatic occurs. Nothing needs to. The system simply becomes less precise.

Ageing is often spoken of as failure, and cancer as rebellion. Both metaphors mislead. Neither process is foreign to the body, and neither is accidental. They arise from the same logic that sustains life in the first place: the imperfect maintenance of complex systems over time. The deeper question, then, is not why bodies fail, but why they fail in such consistent and recognisable ways.

Modern biology has become adept at enumeration. We can catalogue the molecular damage that accumulates with age, the pathways that drift, the cellular behaviours that precede malignancy. These inventories have been indispensable. They have organised fields, guided experiments, and shaped medicine. Yet even the most comprehensive list leaves something unresolved. Description answers what happens. It does not explain why living systems that evolved formidable mechanisms of repair and restraint so reliably come undone.

The answer lies not in error, nor in neglect, but in compromise. Life persists by trade-offs. Evolution does not optimise for longevity; it optimises for sufficiency. It favours traits that promote growth, adaptability, and survival early in life, even when those same traits later undermine stability. Repair is prioritised, but not perfectly. Surveillance exists, but not without blind spots. Restraint is enforced, but never without cost. The result is not decay by design, but fragility by necessity, *vita brevis*, sustained by balance rather than foresight.

This logic was implicit from the moment biology first took history seriously. When Charles Darwin described natural selection, he was careful to limit its reach. Selection acts on immediate advantage, not on distant outcomes; it rewards what works now, not what lasts longest. Multicellular life, with its division of labour, its cooperative tissues, its suppression of individual cellular ambition, is therefore a precarious achievement. It persists only so long as mechanisms of restraint hold. That they eventually loosen is not a failure of selection, but a consequence of how selection works.

This book examines seven recurring patterns through which that loosening occurs. I call them cellular sins, not to suggest intention or moral failure, but to name a persistent excess: growth without adequate restraint, survival without timely exit, response without resolution. These patterns are neither isolated nor accidental. They recur across tissues, across organisms, and across time because they reflect the same underlying tension, between the interests of individual cells and the coherence of the organism they inhabit.

Each sin begins as a virtue. Proliferation enables development and repair. Resistance to death preserves tissue integrity. Sensitivity to nutrients allows organisms to exploit abundance. Inflammation defends against infection. Yet virtues extended beyond their proper context become liabilities. When sustained too long, amplified too strongly, or insufficiently constrained, they erode the collective order they once served. Cells act locally; consequences accumulate globally, *quod licet cellulae, non licet corpori.*

Across biology, function depends on balance rather than maximisation. Cellular processes operate within narrow zones of adequacy: too little activity compromises maintenance and repair, while excess destabilises coordination and structure. Longevity is therefore constrained not by the absence of mechanisms, but by the difficulty of keeping them within this optimal range over time.

What follows can be understood as gradual departures from such biological sufficiency, rather than abrupt failure. This narrow region of adequacy resembles what is often described as a biological Goldilocks zone, not because it is ideal, but because it is intolerant of deviation.

Cancer is the most uncompromising expression of this logic. It is not an invasion from without, but an escalation from within, the consequence of cellular autonomy carried beyond the limits that multicellular life can tolerate. Ageing is slower, quieter, more diffuse. It lacks the drama of malignancy, but not its logic. Both arise from the same biological bargain, differing chiefly in pace and visibility rather than principle.

None of this implies resignation. Biology is malleable, and intervention is possible. Many of the processes discussed here can be delayed, modulated, or partially reversed. But intervention is never neutral. Measures that enhance regeneration may increase the risk of cancer; those that enforce restraint may impair repair. Longevity is not free. Every extension exacts a cost, and those costs must be understood before they are paid.

This book does not promise escape from mortality, nor does it rehearse the familiar optimism of technological salvation. Its aim is more exacting: to explain why living systems unravel as they do, and why preventing one form of decline so often invites another. In tracing these seven cellular sins, we are not indicting biology. We are recognising the price of complexity, the cost of being adaptable, resilient, and alive for longer than evolution ever required.

Once this structure is seen clearly, it becomes difficult to imagine ageing or cancer in any other way.

The argument of this book unfolds in three distinct movements, each addressing the same problem at a different level of resolution.

The first establishes the framework. It identifies seven recurring patterns through which multicellular cooperation erodes with time. These "cellular sins" are not presented as independent mechanisms, nor as moral metaphors, but as characteristic excesses, forms of persistence, growth, response, or competition that become destructive when insufficiently constrained. No single sin is primary. Each gains force through interaction with the others.

The second movement subjects this framework to stress. It steps away from metaphor and examines what can be observed:

historical constraints imposed by multicellularity, population-level incidence of disease, tissue-level ecological change, and measurable shifts in cellular behaviour with age. These sections do not add new claims. They test whether the framework survives contact with evidence drawn from different scales of biology.

The third movement changes perspective. Having established and tested the structure, it asks what kind of knowledge this is, and what follows from possessing it. It does not offer prescriptions or promises. It considers instead what restraint, responsibility, and intervention mean once limits are understood to be structural rather than accidental.

This organisation is deliberate. The argument does not proceed from cause to cure, but from pattern to constraint. What follows is not a catalogue of failures, but an attempt to understand why decline takes the forms it does, and why efforts to correct one excess so often provoke another.

The Seven Cellular Sins

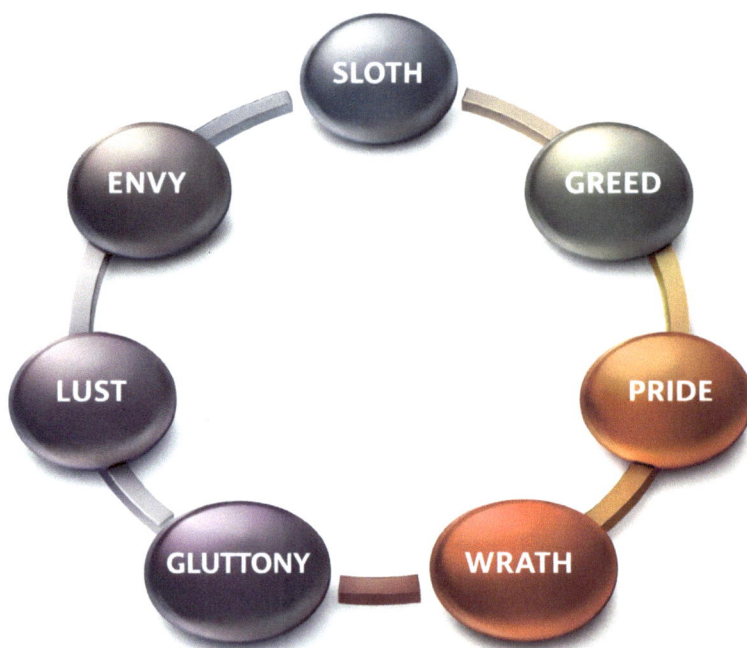

The seven cellular sins represent recurrent excesses that emerge as constraints on multicellular cooperation relax. No single sin is primary. Each amplifies the others, producing ageing and cancer as system-level outcomes rather than isolated failures.

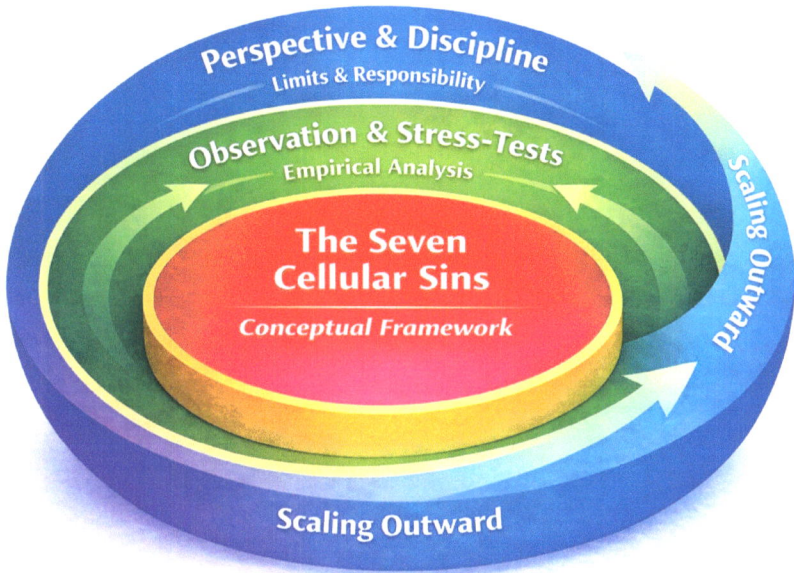

Three analytical rings of the book.

The framework is introduced conceptually through the seven cellular sins (centre), tested against empirical observation across biological scales (middle ring), and finally reframed in terms of limits, responsibility, and discipline (outer ring). The analysis does not progress linearly, but outward in perspective.

Chapter 1

Sloth

Because thou art lukewarm, and neither cold nor hot…
- Revelation 3:16

We encounter this failure without noticing it.
It appears as small delays in repair, minor imperfections
tolerated rather than corrected, and gradual loss of precision that
never triggers alarm.
Only later do we recognise it as a systemic relaxation of
maintenance rather than simple wear.

The Failure to Maintain

Nothing in biology is built to last.

Living systems persist not because they are stable, but because they are continuously repaired. Proteins are replaced, membranes renewed, DNA surveilled, organelles recycled. At every moment, the body expends energy not merely to function, but to remain itself. Life is work, and ageing begins when that work is done less thoroughly than before.

The failure of maintenance is neither sudden nor dramatic. It does not announce itself as a breakdown, but as a marginal loss of fidelity. Repairs are still made, but less completely. Errors are still corrected, but more slowly. Waste is still removed, but less efficiently. The system continues to function, yet it does so with a growing residue of imperfection.

This is the first cellular sin: not indolence, but *underinvestment*. A gradual retreat from the costly labour of upkeep.

Maintenance is expensive. The molecular machinery required to fold proteins correctly, repair damaged DNA, clear dysfunctional organelles, and remove defective cells consumes vast amounts of energy and resources. In early life, that expense is justified. Growth, reproduction, and survival depend on precision. Later, the calculus shifts. Selection no longer strongly rewards perfect repair once reproduction is complete. What remains is a system that continues to operate, but with diminishing insistence on exactness.

This is not a flaw in design. It is the consequence of design under constraint.

At the cellular level, the signs are subtle but cumulative. Proteins misfold and persist. Autophagic clearance slows. DNA lesions accumulate not because repair ceases, but because it becomes incomplete. Mitochondria linger past optimal function. Senescent cells are tolerated rather than removed. None of these changes is catastrophic in isolation. Together, they alter the internal environment in which cells operate.

Ageing, in this sense, is not the accumulation of damage alone, but the accumulation of *unresolved* damage.

The consequences of diminished maintenance are not confined to ageing. Cancer emerges from the same permissive environment. Cells that would once have been eliminated are now allowed to persist. Mutations that would once have been corrected are now tolerated. The gradual slackening of surveillance creates ecological space, and biology abhors a vacuum. Where oversight weakens, selection intensifies.

It is tempting to imagine that maintenance could simply be restored: that ageing reflects a failure of effort rather than a limit of strategy. Yet maintenance systems are already operating near the edge of what is energetically sustainable. Perfect repair is theoretically conceivable but biologically prohibitive. A body that invested indefinitely in flawless maintenance would sacrifice growth, reproduction, and adaptability. It would survive, but it would not compete.

Sloth, then, is not a lapse. It is a negotiated compromise.

Autophagy illustrates this logic with particular clarity. The recycling of intracellular components is essential for cellular health, yet it is also metabolically demanding. In youth, autophagic flux is robust, responsive, and tightly regulated. With age, it declines, not uniformly, but selectively. Some components are cleared efficiently; others persist. The cell becomes cluttered, not with debris alone, but with history.

The same pattern recurs across maintenance systems. DNA repair pathways remain active but incomplete. Proteostasis networks continue to function but lose precision. Immune surveillance persists but becomes tolerant. Each system degrades not because it fails outright, but because it relaxes its standards.

This relaxation has consequences. In tissues with high turnover, diminished maintenance leads to exhaustion. In long-lived tissues, it leads to accumulation. In both cases, the balance shifts from renewal to persistence, a shift that favours cellular survival over tissue coherence.

Cancer exploits this shift mercilessly. Tumour cells thrive in environments where repair is imperfect and clearance delayed. They are not created by sloth, but they are enabled by it. The same permissiveness that allows ageing cells to endure also allows malignant ones to escape elimination. What differs is not the underlying logic, but the tempo.

Attempts to counteract this sin have met with limited success. Enhancing maintenance can delay aspects of ageing, but it often comes at a cost. Increased surveillance may suppress tumours while impairing regeneration. Enhanced clearance may rejuvenate tissues while increasing vulnerability elsewhere. The system resists optimisation because it was never meant to be optimised in this way.

Sloth, then, is the quietest of the sins, and the most foundational. Without it, the others could not proceed. Growth becomes dangerous only when maintenance lags. Persistence becomes pathological only when clearance falters. Disorder gains purchase only when repair grows permissive.

Ageing begins not with rebellion, but with accommodation.

Once maintenance is allowed to relax, everything else follows.

Coda: *Where Maintenance Can Be Nudged - and Where It Cannot*

If the failure of maintenance is the first cellular sin, it is reasonable to ask whether it can be corrected. Experimental biology suggests that elements of cellular maintenance can indeed be strengthened, but never unconditionally, and never without cost.

The clearest evidence concerns autophagy, the intracellular recycling system responsible for the removal of damaged proteins, organelles, and other cellular debris. Declining autophagic flux is now recognised as a hallmark of ageing, and genetic or pharmacological stimulation of autophagy improves healthspan and delays functional decline in multiple model organisms (López-Otín *et al.*, 2023). In mice, enhanced autophagy preserves tissue integrity, improves metabolic function, and attenuates age-associated pathology. In simpler organisms, similar interventions can extend lifespan substantially.

Yet autophagy is not a universal good. Its effects are profoundly context-dependent. While autophagy suppresses tumour initiation by limiting genomic instability and maintaining proteostasis, it can also sustain established tumours by enabling survival under metabolic stress, hypoxia, and therapeutic insult (White, 2015; Kimmelman & White, 2017). In this setting, the same process that protects tissues early in life preserves malignant cells later on. Autophagy, in other words, does not distinguish between cells that ought to persist and those that ought to be eliminated.

This duality is not unique to autophagy. DNA repair mechanisms exhibit similar trade-offs. Enhancing repair capacity can reduce mutation burden and delay certain aspects of ageing, but it may also permit the survival of damaged cells that would otherwise undergo senescence or apoptosis, thereby increasing long-term cancer risk (López-Otín *et al.*, 2023). Perfect fidelity is neither biologically attainable nor necessarily adaptive. Repair systems evolved to be sufficient, not exhaustive.

Even interventions that appear broadly beneficial, caloric restriction, intermittent fasting, or pharmacological mimetics, act by reallocating energy away from growth and towards maintenance. Their effects depend critically on timing, tissue type, and genetic background, and they are accompanied by measurable costs, including impaired wound healing, reduced fertility, or altered immune responses (Madeo *et al.*, 2019). What improves function in one context may compromise it in another.

These findings do not argue against intervention. They argue against simplification. Maintenance can be modulated, but not perfected. The gradual underinvestment in upkeep that characterises ageing reflects not neglect, but the limits imposed by energetic cost and evolutionary priority. Systems that insist on indefinite repair would forfeit adaptability, growth, and reproductive success.

Sloth, then, is not a failure of effort. It is the consequence of a bargain that biology cannot escape. Maintenance can be delayed, sometimes meaningfully so, but it cannot be made free, and it cannot be made absolute.

References:

1. Kimmelman AC, White E.
    Autophagy and tumor metabolism. *Nature Reviews Cancer*.
    2017;17:305–317.
2. López-Otín C, Blasco MA, Partridge L, Serrano M,
    Kroemer G.
    Hallmarks of aging: An expanding universe. *Cell*.
    2023;186:243–278.
3. Madeo F, Carmona-Gutierrez D, Hofer SJ, Kroemer G.
    Caloric restriction mimetics against age-associated
    disease: targets, mechanisms, and therapeutic
    potential. *Nature Reviews Drug Discovery*. 2019;18:645–666.
4. White E.
    The role for autophagy in cancer. *Nature Reviews Cancer*.
    2015;15:683–695.

Chapter 2

Greed

Take heed, and beware of covetousness.
- Luke 12:15

Growth is usually welcomed as a sign of health.
We learn to fear it only when it exceeds context, when
expansion continues despite declining order.
What begins as repair becomes accumulation, and then
liability.

Growth Beyond Restraint

Growth is the great success of biology.

From a single cell, an organism assembles tissues of extraordinary complexity: layered epithelia, branching vasculature, neural networks of improbable density. Growth repairs wounds, replaces losses, and adapts form to function. Without it, life would be brittle. With too much of it, life becomes dangerous.

This is the second cellular sin: growth that continues after its purpose has been served.

Greed, in this context, is not uncontrolled proliferation alone. It is the sustained prioritisation of expansion over maintenance, of synthesis over repair, of immediate gain over long-term coherence. It reflects a system tuned to respond to abundance, nutrients, growth factors, favourable conditions, but insufficiently equipped to disengage once those signals persist.

In early life, such responsiveness is essential. Development demands rapid cell division. Tissues must expand, remodel, and differentiate. Nutrient-sensitive pathways ensure that growth proceeds when resources permit and slows when they do not. The logic is sound. The problem arises when the same logic is extended indefinitely.

With age, growth signals do not simply diminish. They become misaligned.

Nutrient-sensing pathways continue to respond to availability rather than necessity. Cells interpret abundance as instruction. Protein synthesis remains active even as quality control falters. Proliferative cues persist in tissues whose architecture no longer tolerates expansion. What was once adaptive becomes excessive, not because growth accelerates, but because restraint weakens.

This imbalance is subtle. Most cells do not divide recklessly. They respond appropriately to local cues. Yet across tissues, the cumulative effect is a bias: towards synthesis rather than renewal, towards accumulation rather than turnover. Hypertrophy replaces regeneration. Fibrosis replaces flexibility. Structure thickens where it once adapted.

Ageing tissues often grow heavier, denser, stiffer, not because they are healthier, but because growth has outlasted its design.

Cancer represents the extreme end of this continuum. It is growth liberated from context. But it does not emerge from nothing. It exploits pathways already present, already active, already valued. The machinery of proliferation, nutrient uptake, and biosynthesis is not invented by tumours; it is inherited. What cancer adds is persistence, the refusal to yield when growth no longer serves the organism.

The metabolic rewiring seen in malignant cells is often described as pathological. Yet its components are familiar. Increased glucose uptake, enhanced biosynthesis, resistance to nutrient scarcity, these are exaggerations of normal growth programmes. In youth, they build bodies. In age, they burden them. In cancer, they overwhelm them.

Greed is therefore not an anomaly. It is a consequence of success carried too far.

Attempts to restrain this sin have been among the most intensively studied in ageing biology. Reducing growth signals, through dietary restriction, altered nutrient composition, or pharmacological modulation, can delay aspects of ageing and extend lifespan in multiple organisms. These interventions work not by repairing damage directly, but by slowing the rate at which it is created. Less growth means less error, less demand, less strain on maintenance systems already under pressure.

Yet restraint is costly. Suppressing growth impairs wound healing, compromises immune responses, and limits regenerative capacity. In developing or injured tissues, such suppression is harmful. Even in adults, excessive restraint risks fragility. The balance is narrow, and biology guards it jealously.

Cancer therapies that target growth pathways reveal the same tension. Inhibiting proliferation can slow tumours, but it also harms rapidly renewing normal tissues. Blocking nutrient sensing may starve malignant cells, but it may also weaken the host. The machinery of greed is shared. Selectivity is limited.

Ageing, once again, is not the failure of a single process, but the gradual erosion of balance. Growth persists where it should yield. Synthesis outpaces clearance. Expansion replaces renewal. The body becomes larger in parts, but poorer in function.

Greed does not announce itself as excess. It presents as normality sustained too long.

When growth ceases to be checked by maintenance, sloth enables greed. When greed dominates repair, other sins follow. Persistence becomes pride. Response becomes wrath. Survival becomes obsession.

The logic tightens.

Coda: *Constraining Growth, and the Price of Doing So*

If Greed reflects growth extended beyond its appropriate bounds, it follows that restraining growth might slow biological decline. Decades of work across model organisms support this idea, though never without qualification.

The most robust evidence comes from interventions that attenuate anabolic signalling. Reduced activity of insulin and insulin-like growth factor pathways extends lifespan in organisms ranging from nematodes to mammals, an effect first established through genetic manipulation rather than dietary intervention (Kenyon *et al.*, 1993; Holzenberger *et al.*, 2003). These studies demonstrated that diminished growth signalling can prolong life even in the absence of overt caloric deprivation, implicating growth control itself, not simply energy intake, as a determinant of ageing.

Pharmacological inhibition of downstream growth regulators yields similar results. Suppression of mTOR signalling extends lifespan when initiated in adult or even late-life mice, indicating that excessive growth signalling remains detrimental beyond development (Harrison *et al.*, 2009). Importantly, these benefits arise not from enhanced repair, but from reduced biosynthetic demand. Slower growth generates fewer errors and imposes less strain on maintenance systems already compromised by age.

Yet growth pathways are indispensable. Genetic or pharmacological suppression of anabolic signalling impairs wound healing, compromises immune competence, and reduces tolerance to physiological stress. In humans, low circulating IGF-1 levels correlate with reduced cancer risk but are also associated with frailty, sarcopenia, and impaired recovery from illness in older populations (Fontana *et al.*, 2016). Longevity gains are therefore accompanied by vulnerability.

Cancer biology exposes this trade-off with particular clarity. Tumours depend on sustained growth and nutrient uptake, making growth pathways attractive therapeutic targets. Inhibition of these pathways can slow tumour progression, but malignant cells often adapt metabolically, activating alternative nutrient sources or rewiring biosynthesis to escape suppression (Pavlova & Thompson, 2016). Meanwhile, normal proliferative tissues, bone marrow, intestinal epithelium, immune cells, bear the cost of systemic growth restraint.

Dietary strategies aimed at restraining growth reflect the same balance. Protein restriction, amino acid modulation, and fasting regimens can reduce tumour incidence and improve metabolic health in animal models, but their effects depend strongly on age, sex, genetic background, and disease state (Levine *et al.*, 2014). What restrains pathological growth in one context may compromise resilience in another.

These findings point to a central limitation. Growth is not an error layered atop biology; it is one of its organising principles. Attempts to suppress Greed succeed only when applied selectively, transiently, or incompletely. Persistent restraint trades one risk for another.

Greed, like Sloth, can be tempered but not removed. Growth must continue if tissues are to regenerate and organisms are to adapt. The problem is not that growth occurs, but that it persists after the systems that once constrained it have begun to fail.

Constraining growth can slow decline. It cannot, on its own, restore balance.

References:

1. Fontana L, Partridge L, Longo VD.
   Extending healthy life span from yeast to
   humans. *Science*. 2016;351:aad3872.
2. Harrison DE, Strong R, Sharp ZD, Nelson JF, Astle
   CM, Flurkey K, et al.
   Rapamycin fed late in life extends lifespan in genetically
   heterogeneous mice. *Nature*. 2009;460:392–395.
3. Holzenberger M, Dupont J, Ducos B, Leneuve P,
   Géloën A, Even PC, Cervera P, Le Bouc Y.
   IGF-1 receptor regulates lifespan and resistance to
   oxidative stress in mice. *Nature*. 2003;421:182–187.
4. Kenyon C, Chang J, Gensch E, Rudner A, Tabtiang R.
   A *C. elegans* mutant that lives twice as long as wild
   type. *Nature*. 1993;363:305–308.
5. Levine ME, Suarez JA, Brandhorst S, et al.
   Low protein intake is associated with a major reduction
   in IGF-1, cancer, and overall mortality in the 65 and
   younger but not older population. *Cell Metabolism*.
   2014;19:407–417.
6. Pavlova NN, Thompson CB.
   The emerging hallmarks of cancer metabolism. *Cell
   Metabolism*. 2016;23:27–47.

Chapter 3

Pride

Pride goeth before destruction, and an haughty spirit before a fall.
- Proverbs 16:18

Growth is usually welcomed as a sign of health.
We learn to fear it only when it exceeds context, when
expansion continues despite declining order.
What begins as repair becomes accumulation, and then
liability.

The Refusal to Exit

Multicellular life depends as much on death as on growth.

Cells are not meant to persist indefinitely. They are born with roles, occupy niches, and are removed when those roles are complete or when damage renders them unsafe. This turnover is not a failure of biology, but one of its central achievements. Renewal requires exit. Integrity depends on timely disappearance.

The third cellular sin is the erosion of that principle.

Pride, in this context, is not arrogance but *persistence*: the continued survival of cells that no longer serve the collective. It is the refusal to exit when exit would preserve order.

In early life, mechanisms enforcing cellular death are precise and uncompromising. Cells with irreparable damage undergo apoptosis. Cells that divide aberrantly are eliminated. Cells that exhaust their replicative capacity withdraw from the cell cycle. These processes protect tissues from instability and prevent individual advantage from undermining collective function.

With age, these mechanisms do not vanish. They soften.

Cells increasingly evade programmed death, not through dramatic mutation, but through gradual shifts in threshold. Damage that once triggered elimination is tolerated. Stress that once enforced withdrawal becomes survivable. The boundary between persistence and pathology blurs.

One manifestation of this shift is cellular senescence, a state in which cells permanently exit the cell cycle but resist removal. Senescent cells do not divide, but they do not leave. They accumulate, occupy space, alter local signalling, and secrete factors that reshape their environment. What begins as a protective response becomes a burden.

Senescence is often described as a brake on cancer, and in its early deployment it is exactly that. By halting the proliferation of damaged cells, it prevents malignant expansion. Yet by resisting clearance, senescent cells persist beyond their usefulness. Their secretory activity promotes inflammation, disrupts tissue structure, and alters the behaviour of neighbouring cells. The brake remains engaged, but the vehicle deteriorates.

This pattern, protection transformed into liability, recurs across ageing tissues. Cells that should be removed linger. Stem cells that should withdraw persist in a compromised state. Immune cells that should be replaced remain active but dysregulated. The organism becomes crowded with experience, but impoverished in function.

Cancer represents the most extreme form of pride. Malignant cells do not merely persist; they actively evade death. They disable apoptotic pathways, resist immune clearance, and survive conditions that would eliminate normal cells. Yet the strategies they employ are not foreign. They are exaggerations of mechanisms already present, already softened by age.

The refusal to exit is therefore not a sudden betrayal. It is a gradual permissiveness that cancer exploits.

Importantly, pride is not confined to cells that divide. Long-lived, non-proliferative cells also resist removal, even when their function deteriorates. Neurons, muscle fibres, and structural cells persist for decades, accumulating damage that cannot be diluted by division. Their survival is essential, but it is not without cost.

Longevity trades replacement for endurance, and endurance carries history with it.

The immune system reflects this tension acutely. Cells trained by experience persist to provide memory and protection. With age, however, memory crowds out naïveté. The repertoire narrows. Old cells dominate, but they respond poorly to new threats. Persistence becomes rigidity.

Attempts to correct this sin have revealed its delicacy. Removing persistent cells can rejuvenate tissues and restore function, but indiscriminate clearance risks collapse. Cells that resist death do so for reasons, some protective, some adaptive. The challenge lies not in enforcing exit, but in discerning when exit serves the whole.

Ageing, once again, emerges not from a single failure, but from a shift in balance. Survival is favoured over replacement. Persistence over renewal. Memory over flexibility.

Pride does not shout. It settles.

And once cells cease to leave, the conditions for the next sins are established. Persistent cells amplify inflammation. They distort nutrient signalling. They reshape their environment in ways that favour survival over coherence. The body becomes a collection of survivors rather than a system of renewal.

Exit, it turns out, is as essential to life as arrival.

Coda: *Enforcing Exit, and the Risk of Doing So*

If Pride reflects the persistence of cells beyond their usefulness, it is natural to ask whether such cells can be removed. Over the past decade, experimental biology has shown that selective clearance of persistent cells can indeed improve tissue function, but it has also revealed how narrow the margin for success remains.

The clearest evidence concerns cellular senescence. Senescent cells accumulate with age across tissues and contribute to functional decline through altered signalling, extracellular matrix remodelling, and the secretion of pro-inflammatory factors. In mouse models, genetic strategies that eliminate senescent cells improve multiple measures of healthspan and modestly extend lifespan, even when initiated late in life (Baker *et al.*, 2016). These findings established that persistent cells are not merely markers of ageing, but active contributors to it.

Pharmacological approaches have reinforced this conclusion. Compounds now termed *senolytics* selectively induce death in subsets of senescent cells, improving physical function, vascular health, and metabolic parameters in aged mice (Zhu *et al.*, 2015; Xu *et al.*, 2018). Importantly, these benefits arise without wholesale tissue ablation, suggesting that selective enforcement of exit is, in principle, achievable.

Yet senescence is not an error to be erased indiscriminately. Senescent cells play essential roles in wound healing, tissue remodelling, embryonic development, and tumour suppression. Eliminating them too aggressively impairs regeneration and compromises structural integrity. In some contexts, senescent cells are protective precisely because they persist (Demaria *et al.*, 2014). Clearance that is beneficial in aged tissue may be harmful during repair or stress.

Immune surveillance further complicates this balance. Under youthful conditions, senescent and damaged cells are recognised and removed by components of the innate and adaptive immune system. With age, this clearance becomes less efficient. Natural killer cells, macrophages, and cytotoxic T cells retain function, but their coordination falters, allowing persistent cells to accumulate (Ovadya & Krizhanovsky, 2018). The problem is therefore not simply the presence of persistent cells, but the gradual failure of systems designed to remove them.

Cancer again exposes the limits of intervention. Malignant cells evade immune-mediated clearance by exploiting tolerance mechanisms that also protect normal tissues. Therapies that enhance immune killing can restore exit for tumour cells, but they risk autoimmunity, chronic inflammation, or tissue damage when applied systemically. The machinery that enforces cellular death is shared; its activation cannot be cleanly confined.

Human data, though still limited, reinforce caution. Early clinical studies of senolytic agents suggest potential benefits in fibrotic and metabolic disease, but also reveal off-target effects and context-dependent toxicity. The long-term consequences of repeated clearance cycles, particularly in tissues reliant on limited stem cell pools, remain unknown.

These findings point to a central constraint. Persistence is not tolerated because biology has forgotten how to eliminate cells, but because indiscriminate elimination would destabilise tissues. Pride emerges where clearance must be balanced against preservation.

Exit can be enforced, but only selectively, intermittently, and incompletely. Too little clearance permits decline; too much invites collapse.

Pride, like the sins before it, reflects not failure of control, but the cost of restraint applied imperfectly over time.

References:

1. Baker DJ, Wijshake T, Tchkonia T, LeBrasseur NK, Childs BG, van de Sluis B, et al.
   Naturally occurring p16Ink4a-positive cells shorten healthy lifespan. *Nature*. 2016;530:184–189.
2. Demaria M, Ohtani N, Youssef SA, et al.
   An essential role for senescent cells in optimal wound healing through secretion of PDGF-AA. *Developmental Cell*. 2014;31:722–733.
3. Ovadya Y, Krizhanovsky V.
   Strategies targeting cellular senescence. *Nature Reviews Molecular Cell Biology*. 2018;19:577–593.
4. Xu M, Pirtskhalava T, Farr JN, et al.
   Senolytics improve physical function and increase lifespan in old age. *Nature Medicine*. 2018;24:1246–1256.
5. Zhu Y, Tchkonia T, Pirtskhalava T, et al.
   The Achilles' heel of senescent cells: from transcriptome to senolytic drugs. *Aging Cell*. 2015;14:644–658.

Chapter 4

Wrath

The wrath of man worketh not the righteousness of God.
- James 1:20

Inflammation is one of the body's most powerful
protections.
When it fails to resolve, it reshapes tissues even in the
absence of injury.
Defence becomes background condition rather than
response.

When Defence Becomes Damage

Inflammation is among the body's most effective tools.

It isolates injury, recruits repair, and eliminates threats. Without it, wounds would fester and infections would spread unchecked. In its acute form, inflammation is decisive, local, and self-limiting. It arrives with purpose and departs once that purpose is served.

The fourth cellular sin begins when it does not depart.

Wrath, in this context, is not excess force, but *unresolved defence*. A response that remains active after the danger has passed, or that is repeatedly reawakened by signals it was never meant to interpret as threats.

In youth, inflammatory responses are tightly regulated. Damage triggers action; repair restores calm. Immune cells arrive, clear debris, coordinate healing, and withdraw. The tissue returns to equilibrium, often stronger for the experience.

With age, that sequence falters.

Persistent cells, unresolved damage, and accumulated molecular debris generate signals that resemble danger but lack a clear target. The immune system responds as it should, yet finds nothing definitive to eliminate. Activation becomes chronic. Resolution fails. The system remains on alert, expending energy and inflicting collateral damage without achieving closure.

This is inflammation without an enemy.

Low-grade, chronic inflammation is now recognised as a defining feature of ageing tissues. It does not announce itself dramatically.

There is no fever, no swelling, no overt pain. Instead, there is a background hum of immune activity: cytokines elevated slightly above baseline, immune cells lingering where they should transiently pass, signalling pathways partially engaged but never completed.

The effects are cumulative.

Chronic inflammation degrades tissue architecture. It alters extracellular matrices, disrupts stem cell niches, and interferes with regenerative signalling. Cells exposed to inflammatory mediators change their behaviour, becoming more resistant to death, more prone to dysfunction, and less responsive to normal regulatory cues. What was meant to protect the tissue begins to reshape it.

Cancer thrives in such environments.

Tumours do not simply tolerate inflammation; they exploit it. Inflammatory signals promote angiogenesis, support metabolic rewiring, and suppress effective immune surveillance. The same immune mediators that damage normal tissues can be co-opted to sustain malignant growth. Wrath, once unleashed, becomes indiscriminate.

Importantly, chronic inflammation is not caused by a single failure. It emerges from the sins that precede it. Sloth allows damage to accumulate. Greed increases metabolic and biosynthetic stress. Pride permits damaged and senescent cells to persist. Each contributes signals that the immune system is trained to interpret as danger. Wrath is therefore not the origin of decline, but its amplifier.

The immune system itself ages under this burden. Cells repeatedly activated lose precision. Memory dominates over flexibility. Responses become exaggerated yet ineffective. The distinction between self and threat blurs. Autoimmunity and immunodeficiency begin to coexist, a paradox that defines immune ageing.

Attempts to suppress inflammation reveal the limits of correction. Anti-inflammatory interventions can alleviate symptoms and slow certain pathologies, but they also impair host defence. Suppressing immune activity increases vulnerability to infection and malignancy. Enhancing it risks autoimmunity and tissue damage. Once again, the balance is narrow.

Wrath is not a malfunction of immunity. It is immunity operating in an environment it was never designed to police indefinitely.

Ageing tissues increasingly resemble chronic wounds: never acutely injured, never fully healed. The immune system remains engaged, but its efforts are misdirected. Defence becomes background damage. Protection becomes erosion.

This is the paradox of Wrath. The response that once preserved order now accelerates disorder, not because it is too strong, but because it cannot conclude.

And as inflammation reshapes tissues, it alters metabolism, nutrient signalling, and cellular behaviour in ways that prepare the ground for the next sin. Excess fuel accumulates. Signals grow louder. Regulation gives way to saturation.

Wrath does not end the story.

It prepares what follows.

Coda: *Quelling inflammation, and why it rarely stays quelled*

If Wrath reflects defence that fails to resolve, it follows that suppressing inflammation might restore order. Decades of research suggest that inflammatory tone can indeed be lowered, but also that doing so exposes deep biological trade-offs.

Chronic, low-grade inflammation is now recognised as a defining feature of ageing, often termed *inflammaging*. This state arises not from overt infection, but from the cumulative presence of damaged macromolecules, persistent cells, and altered tissue environments that continuously stimulate innate immune pathways (Franceschi *et al.*, 2018). In this setting, immune activation is not misdirected; it is responding appropriately to ambiguous signals that cannot be fully eliminated.

Experimental suppression of inflammatory mediators yields mixed results. Genetic or pharmacological inhibition of key inflammatory pathways reduces tissue damage and delays certain age-associated pathologies in animal models. However, these benefits are often context-dependent and accompanied by increased susceptibility to infection or impaired tissue repair (Furman *et al.*, 2019). Inflammation, once again, proves difficult to disentangle from protection.

The problem lies not solely in activation, but in resolution. Acute inflammation is normally followed by an active process of termination, involving specialised pro-resolving mediators that dampen immune responses and restore tissue equilibrium. With age, this resolution phase becomes less effective. Immune cells persist at sites of minor damage, cytokine production declines incompletely, and tissues remain in a prolonged state of alert (Nathan & Ding, 2010). Defence fades into background injury.

Cancer biology illustrates the consequences starkly. Tumour-associated inflammation supports angiogenesis, promotes immune evasion, and facilitates invasion and metastasis. Yet tumours do not create inflammation de novo; they exploit inflammatory circuits that evolved to manage injury and infection (Grivennikov *et al.*, 2010). Efforts to suppress tumour-promoting inflammation therefore risk impairing the very immune responses required for tumour control.

Clinical experience reinforces this dilemma. Long-term use of broad anti-inflammatory agents reduces the incidence of certain inflammatory diseases and, in specific contexts, lowers cancer risk. At the same time, such interventions increase vulnerability to infection, delay wound healing, and may interfere with effective immune surveillance in ageing individuals (Furman *et al.*, 2019). Narrow targeting improves safety but limits efficacy.

These observations suggest that Wrath is not sustained because inflammation is excessive, but because the conditions that normally allow it to resolve are no longer reliably restored. Persistent debris, altered tissue architecture, and surviving dysfunctional cells continuously reactivate immune pathways that were designed for transient engagement.

Suppressing inflammation without addressing these upstream signals offers only temporary relief. The response quiets, but the provocation remains.

Wrath, then, cannot simply be silenced. It can be modulated, redirected, and occasionally shortened, but only insofar as the environment that sustains it is also altered. Immune restraint without resolution is not balance; it is suppression.

The persistence of inflammation reflects not a failure of immunity, but the cost of maintaining defence in tissues that no longer fully renew themselves.

References:

1. Franceschi C, Garagnani P, Parini P, Giuliani C, Santoro A.
   Inflammaging: a new immune–metabolic viewpoint for age-related diseases. *Nature Reviews Immunology*. 2018;18:576–590.
2. Furman D, Campisi J, Verdin E, et al.
   Chronic inflammation in the etiology of disease across the life span. *Nature Medicine*. 2019;25:1822–1832.
3. Grivennikov SI, Greten FR, Karin M.
   Immunity, inflammation, and cancer. *Cell*. 2010;140:883–899.
4. Nathan C, Ding A.
   Nonresolving inflammation. *Cell*. 2010;140:871–882.

Chapter 5

Gluttony

The full soul loatheth an honeycomb; but to the hungry soul every bitter thing is sweet.
- Proverbs 27:7

Abundance is not inherently harmful.
It becomes so when signals lose contrast and meaning.
In such environments, regulation gives way to tolerance.

Life evolved under conditions of scarcity.

For most of biological history, nutrients were unpredictable, energy expensive, and growth conditional. Cells therefore evolved exquisite sensitivity to availability. Signals that indicated abundance, glucose, amino acids, lipids, growth factors, were interpreted as permission. Permission to grow, to divide, to invest.

Gluttony begins when that permission no longer expires.

In modern organisms, abundance is not episodic but persistent. Nutrients arrive continuously. Signals remain elevated. Pathways designed to pulse instead saturate. Cells do not simply receive more information; they receive *too much of the same information*, too often, for too long.

The result is not constant growth, but confusion.

Metabolic signalling is not binary. It is contextual, integrating nutrient availability with stress, damage, and demand. Under conditions of chronic excess, that integration degrades. Pathways remain active even when their outputs no longer serve the tissue. Feedback weakens. Sensitivity declines. Signals blur into background noise.

This is the fifth cellular sin: abundance sustained beyond its interpretive capacity.

With age, tissues exposed to excess fuel undergo subtle but pervasive changes. Lipids accumulate where they are not designed to be stored. Glucose flux overwhelms regulatory buffers. Mitochondria adapt to surplus by altering dynamics and efficiency, generating by-products that further stress cellular systems. Metabolism shifts from responsiveness to inertia.

Importantly, gluttony is not confined to diet. It includes signalling excess of all kinds. Growth factors persist. Hormonal rhythms flatten. Inflammatory mediators amplify nutrient cues. The cell becomes a crowded room of instructions, none of them urgent enough to be decisive.

Insulin resistance exemplifies this logic. Initially adaptive, protecting cells from overload, resistance becomes pathological when sustained. Cells respond by demanding more signal to achieve the same effect. Circulating levels rise. Noise increases. Precision is lost. The system compensates locally while destabilising globally.

Ageing tissues are particularly vulnerable to this saturation. Maintenance systems already softened by sloth struggle under metabolic load. Persistent cells permitted by pride exploit abundant resources. Inflammatory environments generated by wrath amplify nutrient signalling further. Gluttony does not arise alone; it compounds what has come before.

Cancer again reveals the extreme.

Tumour cells flourish in environments rich in fuel and permissive in signalling. They are not merely consumers of excess; they are adept interpreters of it. Metabolic flexibility allows them to exploit glucose, lipids, and amino acids interchangeably. Where normal cells lose sensitivity, malignant cells refine it. They convert abundance into advantage.

Yet the metabolic rewiring seen in cancer is not alien. It is an intensification of pathways already strained in ageing tissues.

What differs is selectivity. Tumours narrow their focus; ageing tissues lose it.

Attempts to correct metabolic excess have met with partial success. Restricting nutrients, altering composition, or dampening signalling can improve metabolic health and delay disease. But excess is not eliminated so much as displaced. Cells adapt. Pathways reroute. The burden shifts rather than disappears.

Moreover, metabolic restraint is not free. Energy limitation compromises immune defence, repair, and resilience to stress. In older organisms, the margin for restriction narrows. What sharpens signalling in youth may blunt it later.

Gluttony, then, is not indulgence but saturation. It reflects systems designed for scarcity operating under relentless abundance. Signals persist beyond meaning. Pathways activate without purpose. Decisions become probabilistic rather than precise.

As signalling degrades, cells increasingly rely on persistence rather than regulation. Survival becomes decoupled from function. The stage is set for the next sin: **Lust**, the pursuit of immortality at the cellular level, even as coherence collapses.

Abundance does not guarantee health.

It guarantees complexity.

And complexity, sustained without restraint, exacts its price.

Coda: *Tuning metabolism, and the limits of precision*

If Gluttony reflects metabolic signalling overwhelmed by persistent abundance, then restoring sensitivity appears an obvious remedy. Experimental biology confirms that metabolic signals can be sharpened, but also reveals why this sharpening is inherently constrained.

Chronic nutrient excess disrupts signalling not merely by increasing substrate availability, but by degrading responsiveness. Sustained exposure to glucose, lipids, and amino acids induces adaptive resistance in key pathways, particularly those governing insulin and growth factor signalling. This resistance initially protects cells from overload, but at the cost of signal fidelity (Shulman, 2014). What begins as insulation becomes insensitivity.

Mitochondria adapt to this excess in similarly ambivalent ways. In nutrient-rich environments, mitochondrial dynamics shift to accommodate surplus fuel, altering efficiency, redox balance, and reactive by-product generation. These adaptations preserve short-term function but increase oxidative and metabolic stress over time, further impairing signalling accuracy (Nicholls & Ferguson, 2013). Energy continues to flow, but information is lost.

Interventions that reduce metabolic load can partially restore responsiveness. Weight loss, altered nutrient composition, and improved insulin sensitivity enhance signalling precision in both animal models and humans. Yet these gains are often transient. Cells recalibrate. Pathways adjust their thresholds. The system settles into a new equilibrium rather than reverting to its original state (Samuel & Shulman, 2016).

Cancer biology again exposes the boundary conditions. Tumour cells thrive in metabolically noisy environments by narrowing their interpretive range. They upregulate nutrient transporters, rewire central carbon metabolism, and exploit alternative substrates to maintain growth even as normal cells lose responsiveness (DeBerardinis & Chandel, 2016). Metabolic excess therefore widens the gap between malignant selectivity and normal cellular confusion.

Attempts to pharmacologically dampen metabolic signalling encounter the same dilemma. Agents that improve insulin sensitivity or suppress anabolic signalling enhance metabolic control in some tissues while impairing it in others. Systemic interventions cannot distinguish between excess that degrades function and abundance that supports repair, immunity, or regeneration. Selectivity remains limited.

These findings point to a deeper constraint. Metabolism is not a simple supply chain; it is an information-processing system. Signals gain meaning only when they fluctuate against a background of scarcity. When abundance becomes continuous, signalling saturates, feedback erodes, and precision declines. Correction becomes a matter of degree, not restoration.

Gluttony, then, cannot be undone by subtraction alone. Removing excess improves clarity only up to a point. Beyond that, the system resists refinement because it has already adapted to noise.

Metabolic restraint may slow the drift. It cannot fully recover the interpretive sharpness of systems designed for a world in which abundance was rare.

References:

1. DeBerardinis RJ, Chandel NS.
    Fundamentals of cancer metabolism. *Science Advances*.
    2016;2:e1600200.
2. Nicholls DG, Ferguson SJ.
    Bioenergetics 4. 4th ed. London: Academic Press; 2013.
3. Samuel VT, Shulman GI.
    The pathogenesis of insulin resistance: integrating
    signaling pathways and substrate flux. *Cell*.
    2016;168:852–871.
4. Shulman GI.
    Ectopic fat in insulin resistance, dyslipidemia, and
    cardiometabolic disease. *Journal of Clinical Investigation*.
    2014;124:463–468.

Chapter 6

Lust

The eye is not satisfied with seeing, nor the ear filled with hearing."
- Ecclesiastes 1:8

Renewal requires division, but division cannot be endless.
When contribution is extended unevenly, continuity is
preserved at the cost of diversity.
Persistence replaces succession.

The Pursuit of Immortality

Every cell carries a limit.

Division is not indefinite. Replication erodes structure, accumulates error, and exacts a cost. To persist as a coherent organism, multicellular life enforces boundaries: cells divide only so many times, withdraw when their task is complete, and relinquish the future to their successors. Renewal depends on turnover; longevity depends on replacement.

The sixth cellular sin begins when those limits are resisted.

Lust, in this context, is not desire but *persistence beyond design*. The activation of mechanisms that allow cells to continue when continuation no longer serves the whole.

At the heart of this impulse lies a simple problem. DNA replication is imperfect, and chromosomes shorten with each division. Left unchecked, this erosion would exhaust proliferative tissues. To counter it, biology evolved mechanisms of renewal, most notably the maintenance of telomeres, the repetitive sequences that cap chromosome ends and preserve genomic integrity during replication.

In development and early life, these mechanisms are tightly regulated. They allow stem cells to replenish tissues, immune cells to expand when needed, and germ cells to transmit continuity across generations. Immortality, in this sense, is not pathological. It is essential, but carefully constrained.

With age, that constraint weakens.

Cells under stress, exposed to damage, inflammation, or metabolic excess, increasingly activate pathways that favour survival and continued division. Telomere maintenance is re-engaged. Cell-cycle checkpoints soften. Replicative limits are deferred. What was once reserved for regeneration becomes available to persistence.

This shift is subtle. Most cells do not become immortal. But the threshold for continued survival lowers. Cells that should exit linger. Clones that should extinguish persist. The balance tips from renewal towards endurance.

Cancer represents the most explicit expression of this sin. Malignant cells do not merely evade death; they secure the means to divide indefinitely. By reactivating telomere maintenance pathways or adopting alternative strategies to preserve chromosome ends, they escape replicative exhaustion. Division becomes uncoupled from consequence.

Yet this is not a foreign innovation. The machinery cancer exploits is inherited. It is the same machinery that sustains stem cell compartments and enables tissue repair. What cancer adds is insistence, persistence without regard for architecture or function.

Ageing tissues create fertile ground for this insistence. Maintenance has softened. Growth signals persist. Clearance falters. Inflammatory cues reinforce survival. Under these conditions, cells that acquire the capacity to continue gain a disproportionate advantage. Lust does not appear suddenly; it is selected.

Importantly, the pursuit of cellular immortality carries costs even outside cancer. Long-lived cells accumulate history. Errors persist. Damage is retained rather than diluted. Endurance replaces renewal. Tissues become older not merely because cells age, but because they fail to be replaced.

Nowhere is this tension more apparent than in stem cell compartments. These cells must balance self-renewal with restraint. Too little persistence leads to exhaustion; too much invites instability. With age, this balance shifts. Stem cells persist, but their output declines. Clonal dominance emerges. Diversity narrows. The system survives, but it becomes brittle.

The immune system again offers a revealing example. Long-lived memory cells persist to provide protection, but their accumulation constrains responsiveness. Immortality, distributed unevenly, reshapes the repertoire. What is remembered crowds out what is possible.

Attempts to intervene in this sin reveal its danger. Enhancing telomere maintenance can preserve proliferative capacity, but it also increases genomic instability and cancer risk. Suppressing immortality pathways may limit malignancy, but at the cost of tissue renewal and immune competence. The mechanisms that allow life to continue cannot be cleanly separated from those that allow it to unravel.

Lust, then, is not an excess of life, but a misplacement of it. Persistence replaces succession. Continuation supplants renewal. The future is claimed by cells unwilling to relinquish the present.

This is the deepest tension in multicellular existence. Life must persist, but not *as itself*. Continuity requires surrender.

When cells refuse that surrender, they do not merely survive. They alter the conditions for everything around them. Competition intensifies. Cooperation frays. Selection shifts inward.

The final sin waits there.

Coda: *Extending continuation, and the cost of doing so*

If Lust reflects the cellular refusal of finitude, then limiting or reversing that refusal appears an obvious intervention. Experimental biology confirms that replicative limits can be manipulated, but it also reveals why doing so rarely favours the organism as a whole.

The most direct approach targets telomere maintenance. Telomere shortening imposes a replicative boundary on dividing cells, and restoring telomerase activity preserves proliferative capacity. In experimental systems, enforced telomerase expression delays tissue degeneration and improves regenerative function, particularly in stem cell compartments (Jaskelioff *et al.*, 2011). These findings demonstrate that replicative ageing is, at least in part, malleable.

Yet the same studies reveal a constraint. Telomerase activation increases genomic instability and markedly elevates cancer risk unless paired with stringent tumour-suppressive mechanisms. Telomeres protect chromosomes, but they also shield cells from the consequences of damage. Extending replicative lifespan therefore preserves not only function, but error.

Stem cell biology exposes this trade-off with particular clarity. Adult stem cells rely on limited self-renewal to maintain tissue integrity while avoiding clonal dominance. With age, stem cell pools persist, but their output declines and clonal diversity contracts. Long-lived clones increasingly dominate, reducing adaptability and increasing vulnerability to transformation (Beerman & Rossi, 2015). Persistence secures continuity at the expense of flexibility.

Attempts to suppress immortality pathways encounter the inverse problem. Inhibiting telomerase or enforcing replicative limits can reduce tumour growth and constrain malignant persistence. However, such strategies also impair tissue renewal, accelerate stem cell exhaustion, and compromise immune competence (Shay & Wright, 2019). The boundary between limiting cancer and inducing degeneration is narrow.

Cancer cells again illuminate the underlying logic. Most tumours activate telomerase or alternative telomere-lengthening mechanisms to escape replicative exhaustion. Blocking these pathways can restrict tumour growth, but malignant cells often adapt by slowing division, entering quiescence, or selecting for clones that tolerate genomic instability (Cesare & Reddel, 2010). Immortality is resisted because it is advantageous, but it is never achieved without cost.

Human evidence reinforces caution. Individuals with inherited telomere maintenance disorders experience premature tissue failure, immunodeficiency, and stem cell exhaustion. Conversely, conditions associated with excessive telomere maintenance correlate with increased cancer risk. Both extremes reveal the same principle: replicative limits are neither arbitrary nor negotiable without consequence (Armanios & Blackburn, 2012).

These findings point to a deeper constraint. Cellular immortality is not pathological because it is rare, but because it is misallocated. Mechanisms that preserve continuity in select compartments become destabilising when broadly reactivated. Life extends itself by passing forward, not by lingering.

Lust, then, cannot be cured by enforcing limits alone, nor by abolishing them. Replicative boundaries must exist, but they must also yield where renewal is required. The organism survives not by keeping its cells forever, but by ensuring that no cell does.

Continuation without succession is not longevity.

It is congestion.

References:

1. Armanios M, Blackburn EH.
   The telomere syndromes. *Nature Reviews Genetics*.
   2012;13:693–704.
2. Beerman I, Rossi DJ.
   Epigenetic regulation of hematopoietic stem cell
   aging. *Annual Review of Cell and Developmental Biology*.
   2015;31:555–584.
3. Cesare AJ, Reddel RR.
   Alternative lengthening of telomeres: models,
   mechanisms and implications. *Nature Reviews Genetics*.
   2010;11:319–330.
4. Jaskelioff M, Muller FL, Paik JH, et al.
   Telomerase reactivation reverses tissue degeneration in
   aged telomerase-deficient mice. *Nature*. 2011;469:102–
   106.
5. Shay JW, Wright WE.
   Telomeres and telomerase in normal and cancer stem
   cells. *Nature Reviews Drug Discovery*. 2019;18:565–584.

Chapter 7

Envy

Where envying and strife is, there is confusion.
- James 3:16

Competition is essential between organisms and dangerous
within them.
When it re-emerges among cells, cooperation erodes quietly
before it collapses visibly.
Cancer is only the final expression of this return.

Multicellular life is a truce.

From its earliest origins, it required individual cells to surrender something fundamental: the freedom to compete. Growth became regulated. Resources were shared. Reproduction was restricted to a privileged few. In return, cells gained protection, longevity, and access to environments no single cell could sustain alone.

That truce is not enforced once and for all. It must be renewed continuously.

The final cellular sin arises when it is not.

Envy, in this context, is not resentment but *relative advantage*: the success of one cell at the expense of its neighbours. It is the re-emergence of Darwinian competition within a system built to suppress it.

In youth, multicellular order is robust. Maintenance systems function efficiently. Growth is constrained. Damaged cells exit. Immune surveillance is vigilant. Differences between cells exist, but they are contained. Selection operates weakly, if at all, within tissues.

With age, that containment erodes.

Sloth allows damage to accumulate unevenly. Greed amplifies growth signals differentially. Pride permits persistent cells to linger. Wrath reshapes tissue environments unpredictably. Gluttony saturates signalling pathways. Lust extends the lifespan of cells already advantaged. Together, these changes create heterogeneity, and heterogeneity invites selection.

Cells are no longer equivalent.

Some divide slightly faster. Some resist death more effectively. Some exploit nutrients more efficiently. Some tolerate stress better. None of these advantages is dramatic. None would matter in a tightly regulated system. But regulation has softened. Clearance has faltered. Constraints have loosened.

Selection resumes.

This is the moment when Envy becomes decisive.

Clonal expansion is not an aberration. It is the natural consequence of variation under relaxed control. Cells with even marginal advantages begin to outcompete their neighbours. Their descendants occupy disproportionate space. Diversity contracts. The tissue becomes dominated not by the best-functioning cells, but by the most persistent ones.

Importantly, this competition is not driven by malice or intent. It is driven by opportunity.

Cancer is the most visible outcome of this process, but it is not the only one. Ageing tissues across the body exhibit clonal dominance without overt malignancy. Blood, skin, intestine, liver, all show evidence of expanding clones that carry growth or survival advantages. Most never form tumours. All reshape the tissue in which they reside.

Envy therefore precedes cancer. Malignancy is merely its most extreme expression.

Multicellular cooperation depends on suppressing within-body evolution. Ageing weakens that suppression. The body becomes an ecosystem rather than a collective. Cells adapt to local conditions rather than global needs. What benefits the clone undermines the organism.

This is the deepest reason ageing and cancer are inseparable.

It is tempting to imagine that tighter control could restore cooperation, that better surveillance, stricter repair, or more aggressive clearance might eliminate internal competition. But perfect suppression of selection is neither achievable nor desirable. Variation enables repair. Adaptation supports survival. The same flexibility that allows tissues to respond to injury also permits divergence.

Life cannot abolish evolution without abolishing itself.

The tragedy is not that cells compete, but that they compete *late*, in environments shaped by decades of compromise. Selection acts on systems already strained. Advantage accrues to persistence rather than performance. The winners are not the most cooperative cells, but the least constrained ones.

Once Envy takes hold, the remaining sins reinforce it. Persistent clones amplify inflammation. Metabolic rewiring favours their survival. Clearance becomes increasingly selective. The tissue tilts further towards dominance and away from diversity.

The end state is not chaos, but order of a different kind: simplified, rigid, and fragile.

Ageing, then, is not simply wear and tear. It is the gradual return of evolutionary dynamics to a system that evolved to exclude them. Cancer is not a foreign invasion, but the final failure of the truce.

This does not render intervention futile. It clarifies its limits.

The goal cannot be to eliminate cellular competition entirely. It can only be to delay its resurgence, to preserve cooperation for as long as possible, and to understand the costs of doing so. Longevity is not achieved by perfecting cells, but by sustaining their willingness to yield.

The body holds together only so long as its parts agree to lose.

When that agreement falters, the logic of evolution resumes, patiently, relentlessly, without regard for the whole.

That is why we fall apart.

Coda: *Suppressing competition, and why it always returns*

If Envy reflects the re-emergence of selection within ageing tissues, it is natural to ask whether this internal competition can be restrained. Over the past decade, human data have made clear that it can be delayed, but not abolished.

The most compelling evidence comes from clonal haematopoiesis. Large-scale sequencing studies have shown that, with age, blood becomes increasingly dominated by expanding clones carrying mutations that confer subtle growth or survival advantages. These clones often arise decades before overt disease and are detectable in individuals with no clinical abnormalities (Genovese *et al.*, 2014; Jaiswal *et al.*, 2014). Their expansion reflects not malignancy, but selection operating in a permissive environment.

Similar patterns are now evident across solid tissues. In skin, oesophagus, intestine, and liver, normal ageing is accompanied by widespread clonal mosaicism. Cells carrying mutations that enhance proliferation or stress tolerance gradually displace their neighbours, even while maintaining apparently normal tissue architecture (Martincorena *et al.*, 2015). These clones do not necessarily progress to cancer, but they alter tissue dynamics nonetheless. Competition has resumed.

Crucially, these findings demonstrate that Envy does not require transformation. Selection emerges whenever variation coincides with relaxed constraint. Ageing tissues supply both.

Attempts to suppress this competition encounter a paradox. Enhancing surveillance and clearance can limit clonal dominance, but only by imposing stronger selective pressures. Cells that evade detection or tolerate stress are favoured. Suppression accelerates refinement. Selection sharpens.

Conversely, relaxing control to preserve diversity invites expansion of the most persistent clones. The system drifts towards dominance either way. The difference lies in tempo, not outcome.

Cancer prevention strategies reflect this dilemma. Interventions that reduce mutation rate or promote cell turnover delay clonal expansion, but cannot eliminate it. Even in tissues with robust renewal, selection operates whenever differences arise. Perfect uniformity is unattainable in living systems that must adapt, repair, and respond to stress.

Evolutionary theory offers no escape clause here. Selection is not a process that can be switched off locally while preserved globally. Multicellular life suppresses it temporarily, imperfectly, and at cost. Ageing represents the gradual repayment of that cost.

What changes with time is not the presence of selection, but its direction. In youth, selection favours cooperation, fidelity, and restraint. In ageing tissues, it favours persistence, resistance, and autonomy. The environment has shifted. The winners change accordingly.

These observations do not imply inevitability in a nihilistic sense. Rates matter. Context matters. Interventions that preserve tissue integrity, reduce damage, and maintain clearance can postpone the resurgence of competition. But they cannot repeal it. So long as cells differ, and so long as constraints are finite, Envy will reappear.

The deepest limit of longevity is therefore not damage alone, nor metabolism, nor inflammation, nor even immortality mechanisms. It is the return of evolution to the interior of the body.

We do not age because selection fails.

We age because, eventually, it succeeds.

References:

1. Genovese G, Kähler AK, Handsaker RE, et al.
    Clonal hematopoiesis and blood-cancer risk inferred
    from blood DNA sequence. *New England Journal of
    Medicine*. 2014;371:2477–2487.
2. Greaves M.
    Evolutionary determinants of cancer. *Nature Reviews
    Cancer*. 2015;15:531–542.
3. Jaiswal S, Fontanillas P, Flannick J, et al.
    Age-related clonal hematopoiesis associated with adverse
    outcomes. *New England Journal of Medicine*.
    2014;371:2488–2498.
4. Martincorena I, Roshan A, Gerstung M, et al.
    High burden and pervasive positive selection of somatic
    mutations in normal human skin. *Science*. 2015;348:880–
    886.

The Dial of Cellular Excess

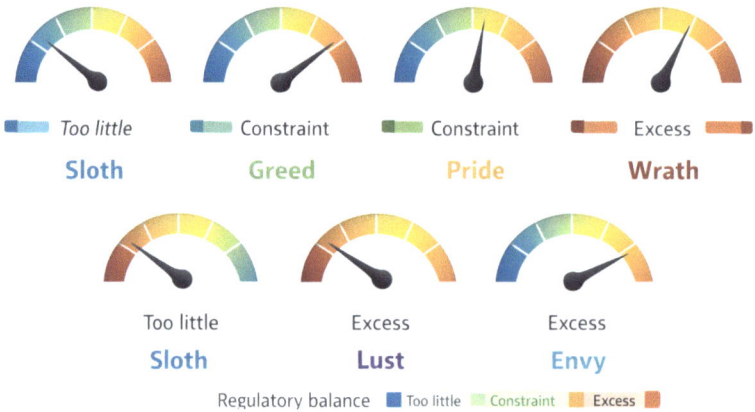

The Dial of Cellular Excess.
Core cellular processes underlying ageing and cancer operate within a narrow zone of restraint. Each dial represents a regulatory continuum, with insufficient activity at one extreme and pathological excess at the other. The seven cellular sins correspond to distinct ways in which this balance is lost over time, reflecting failures of constraint rather than isolated defects.

After the Sins

The language of sin implies the possibility of redemption. Biology offers no such guarantee.

Ageing is often framed as accumulation: damage added to damage, error layered upon error. Cancer is framed as interruption, a sudden deviation from an otherwise orderly decline. Both descriptions mislead. What this book has argued is something more constrained, and more demanding.

Ageing and cancer arise not from isolated failures, but from the gradual unravelling of a truce.

Multicellular life depends on cooperation among cells that would otherwise compete. That cooperation is sustained through maintenance, restraint, clearance, resolution, and renewal, processes that are costly, imperfect, and unevenly enforced. For a time, they hold. Eventually, they loosen. Not because they are neglected, but because they were never designed to operate indefinitely.

The seven cellular sins traced in these pages are not independent flaws. They are different expressions of the same compromise. Maintenance softens. Growth persists. Exit fails. Defence lingers. Signals saturate. Continuation replaces succession. Variation meets relaxed constraint. Selection resumes.

None of this reflects malice or error. Cells do not betray the body. They obey rules shaped by evolution, rules that favour early success over late stability, sufficiency over perfection, adaptability over endurance. What we call decline is the long shadow of those priorities.

Cancer is not an exception to this logic. It is its most uncompromising outcome. It appears when the balance between cooperation and competition finally tips inward, when cellular advantage is no longer reliably subordinated to collective need. Ageing prepares the ground; cancer exploits it. The difference is not one of principle, but of pace.

This perspective carries consequences.

It means that no single intervention can redeem ageing without disturbing something else. Enhancing maintenance strains regeneration. Suppressing growth invites fragility. Enforcing exit risks collapse. Silencing inflammation compromises defence. Sharpening metabolic control reduces resilience. Limiting immortality accelerates exhaustion. Suppressing competition intensifies selection.

Biology resists redemption because it is not an engineering problem. It is a history.

This does not render intervention futile. Rates can be altered. Trajectories can be bent. Periods of function can be extended. Suffering can be reduced. But expectation must be disciplined. Longevity is not achieved by perfecting cells, but by sustaining cooperation among imperfect ones, and cooperation has limits.

To understand ageing, then, is not to uncover a hidden programme of decay, nor to identify a single master switch. It is to recognise the structure of the constraint under which complex life persists. The body does not fall apart because it loses control of its parts, but because control itself becomes increasingly expensive to maintain.

After the sins, there is no absolution.

There is only understanding.

And that understanding clarifies what is possible, what is costly, and what will always resist repair.

We fall apart not because biology is careless, but because it is exacting.

And because, in the end, even cooperation must yield to time.

Part II - Observations and Stress-Tests

On Observation

The chapters that follow differ in purpose from those that precede them.

Part I described a set of recurring patterns through which ageing and cancer emerge from the same underlying logic. Part II asks whether those patterns withstand scrutiny when viewed from other angles. It does not seek corroboration through accumulation, but through tension.

The sections that follow are therefore observational rather than propositional. They draw on historical constraint, epidemiology, tissue ecology, and measurable cellular behaviour to test whether the framework remains coherent across scale and context. Where the argument strains, that strain is instructive. Where it holds, it does so provisionally.

No single observation here is decisive. Taken together, they ask a simpler question: whether the seven sins describe a real structure in living systems, or merely a convenient metaphor. If the latter, the framework should collapse under pressure. If the former, its limits should become visible.

What follows is not evidence in support of an idea, but exposure of an idea to reality.

Multicellularity as a Temporary Truce

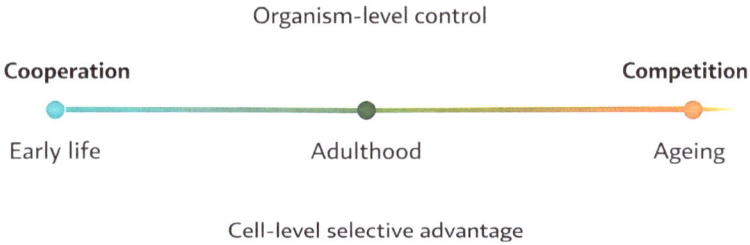

Organism-level control

Cooperation   **Competition**

Early life   Adulthood   Ageing

Cell-level selective advantage

Multicellular organisms persist by suppressing cellular self-interest. This suppression is strongest early in life and gradually weakens with age, allowing cell-level selection to re-emerge. Ageing reflects the progressive relaxation of this constraint.

Part II - Observations and Stress-Tests

II.0 Before the Truce

The Origins of multicellular restraint

Multicellular life did not begin with harmony. It began with constraint.

The transition from unicellular existence to organised multicellularity required a fundamental reordering of evolutionary priorities. Single cells prosper by competing. They divide when resources permit, exploit local advantage, and persist so long as conditions allow. Selection acts directly and continuously. Success is measured locally and immediately.

Multicellular organisms required the suppression of this logic.

For cells to cooperate, division had to be regulated, resources shared, and reproduction restricted. Individual advantage became subordinate to collective function. Cells that might otherwise outcompete their neighbours were constrained by developmental programmes, growth controls, and mechanisms enforcing exit. In evolutionary terms, selection was displaced upward, from cells to organisms.

This displacement was never absolute.

Even in the earliest multicellular lineages, cooperation depended on active enforcement. Developmental patterning, lineage commitment, and programmed cell death did not emerge to optimise longevity, but to stabilise form. These mechanisms limited cellular autonomy sufficiently to permit complexity, but they did not abolish the underlying capacity for competition. They merely held it in check.

Ageing reflects the cost of maintaining that check over time.

From its inception, multicellular life faced a structural dilemma. The same flexibility that allows tissues to grow, adapt, and repair also permits variation. The same persistence that preserves function allows error to accumulate. The same mechanisms that suppress competition early in life must relax to allow renewal and response. There is no static solution.

Early tumour suppression strategies reflect this balance. Limiting cell division reduces the opportunity for error, but impairs growth and repair. Permitting proliferation enables development and regeneration, but increases risk. Organisms evolved layered safeguards, checkpoints, surveillance, clearance, rather than absolute barriers. These safeguards are effective early, costly to maintain, and imperfect by design.

Ageing was not selected against because it was not a discrete trait. It emerged as a consequence of prioritising early-life success in environments where late-life survival exerted little evolutionary pressure. Maintenance systems evolved to be sufficient, not exhaustive. Repair pathways correct common insults, not all possible failures. The organism survives long enough to reproduce; what follows is secondary.

Cancer, viewed in this context, is not an evolutionary oversight. It is the predictable failure mode of a system that trades long-term stability for adaptability and early reproductive success. Tumour suppression delays but does not eliminate risk. The longer the truce is enforced, the greater the cost of enforcing it further.

Importantly, the suppression of selection within organisms was never uniform across tissues. Renewal rates differ. Exposure to damage varies. Stem cell hierarchies impose structure, but also create bottlenecks where advantage can accumulate. These asymmetries are not flaws; they are functional compromises. Over time, however, they shape where competition is most likely to re-emerge.

The history of multicellularity therefore does not point towards a lost ideal of perfect cooperation. It reveals a system perpetually negotiating restraint. Control is applied where it yields the greatest immediate benefit and relaxed where flexibility is required. The balance shifts across development, adulthood, and ageing.

This historical perspective clarifies a key point that recurs throughout this book: the return of competition in ageing tissues is not a reversal of evolution, but its continuation under altered constraints. Selection never disappears. It is deferred, redirected, and partially suppressed, and then, gradually, allowed back in.

Before the truce, cells competed openly.
After it, they cooperate conditionally.

Ageing marks the slow renegotiation of those conditions.

Part II - Observations and Stress-Tests

II.1 Sloth Observed

Maintenance failure as the permissive substrate

Across ageing tissues, decline in maintenance capacity is neither uniform nor catastrophic. It is incremental, uneven, and cumulative. Its most consequential feature is not the appearance of damage itself, but the gradual reduction in the efficiency with which damage is detected, corrected, or cleared.

This reduction is observable across molecular, cellular, and tissue scales.

At the molecular level, ageing cells exhibit increased burdens of damaged proteins, lipids, and nucleic acids. Repair pathways remain operative, but their throughput declines. Errors that would once have been corrected promptly persist longer. Quality control becomes selective rather than comprehensive. The system shifts from prevention to tolerance.

At the cellular level, this manifests as heterogeneity. Some cells maintain effective repair and clearance, others accumulate damage more rapidly. Differences in proteostasis, DNA repair efficiency, and organelle turnover widen with age. Uniformity erodes. Variation increases.

At the tissue level, these differences matter.

Tissues depend on coordinated maintenance to preserve architecture and function. When maintenance becomes uneven, local failures accumulate without immediate consequence. Cells continue to function, but with compromised fidelity. Structures persist, but with latent fragility. Decline proceeds without triggering acute failure.

This pattern distinguishes ageing from injury. In injury, damage exceeds capacity and provokes response. In ageing, damage accumulates below threshold. The system adapts by accommodating imperfection rather than correcting it.

Crucially, maintenance failure does not act alone. It creates permissive conditions for the sins that follow.

Reduced quality control allows damaged but viable cells to persist. Clearance thresholds rise. Cells that would previously have exited remain active. Growth signals encounter weakened restraint. Inflammatory stimuli accumulate. Metabolic stress is tolerated rather than resolved. Maintenance failure does not dictate outcome; it alters the field in which outcomes are selected.

From an ecological perspective, declining maintenance widens the distribution of cellular states within tissues. Some cells remain robust, others marginal. This heterogeneity increases the opportunity for relative advantage to matter. Selection becomes possible not because cells seek dominance, but because constraints no longer equalise their differences.

Importantly, maintenance systems are costly. High-fidelity repair and clearance require energy, coordination, and redundancy. Evolution favoured sufficiency over excess. Repair pathways correct common insults effectively, but rare or cumulative damage is managed imperfectly. With time, the cost of comprehensive maintenance exceeds its evolutionary benefit.

This trade-off is visible in comparative biology. Species with extended lifespans tend to invest more heavily in maintenance, but none achieves completeness. Even long-lived organisms accumulate damage; they simply do so more slowly. Maintenance delays decline; it does not abolish it.

Attempts to augment maintenance experimentally illustrate the same limit. Enhancing repair pathways or clearance mechanisms can reduce damage burden and improve function in specific contexts. Yet such interventions often redistribute rather than eliminate cost. Increased maintenance demands energy, alters growth dynamics, or interferes with adaptability. Gains are context-dependent and finite.

What ageing reveals, therefore, is not neglect of maintenance, but its bounded nature. Systems continue to function because they tolerate imperfection. Decline accelerates when tolerance replaces correction as the dominant strategy.

Sloth observed is not inactivity. It is restraint applied unevenly and increasingly reluctantly as cost rises.

Maintenance does not fail suddenly.

It thins.

And in thinning, it permits everything that follows.

Part II - Observations and Stress-Tests

II.2 Greed Observed

Growth signalling under weakened maintenance

Growth signalling does not cease with age. In many tissues, it persists, adapts, and in some contexts intensifies. What changes is not the presence of growth cues, but the capacity of maintenance systems to regulate their consequences.

In youthful tissues, growth signalling is tightly coupled to need. Nutrient availability, growth factors, and mechanical cues converge on pathways that promote proliferation or hypertrophy when repair or expansion is required. These signals are transient and reversible. Maintenance and quality control scale alongside growth, preserving structure.

With age, this coupling loosens.

Growth-associated pathways remain responsive to abundance and stress, but maintenance capacity declines. Biosynthesis proceeds under conditions in which error correction, proteostasis, and organelle turnover are less efficient. The balance between construction and repair shifts towards accumulation rather than renewal.

This shift is observable across tissues.

In muscle, growth signalling persists despite declining regenerative capacity, leading to hypertrophy without proportional functional gain. In liver and adipose tissue, anabolic pathways remain active even as lipid handling and detoxification become impaired. In vascular tissues, growth cues contribute to thickening and stiffness rather than adaptation. Growth continues, but its outputs change.

At the cellular level, sustained growth signalling under weakened maintenance increases the burden of misfolded proteins, damaged organelles, and replication stress. Cells tolerate these burdens rather than resolving them. Quality thresholds rise. What would once have triggered exit now persists.

This tolerance creates selective asymmetry.

Cells that can sustain growth under compromised maintenance gain relative advantage. Those that maintain tighter coupling between growth and repair are disadvantaged by higher cost. Selection begins to favour persistence under load rather than precision under control.

Importantly, growth signalling itself adapts. Feedback mechanisms blunt sensitivity to avoid catastrophic overload, but this adaptation reduces discrimination. Signals persist at intermediate levels, sufficient to sustain anabolic activity but insufficient to enforce restraint. Growth becomes habitual rather than conditional.

Cancer again illustrates the extreme.

Malignant cells exploit weakened maintenance by sustaining growth while bypassing quality control. They tolerate replication stress, misfolded proteins, and metabolic imbalance through adaptive rewiring. Yet this exploitation builds upon conditions already present in ageing tissues. The difference is not the presence of growth signalling, but the refusal to moderate it.

From a population perspective, tissues exposed to persistent growth cues under declining maintenance show increased heterogeneity. Some cells slow and withdraw. Others adapt and persist. This divergence expands the space in which selection operates.

Interventions aimed at suppressing growth signalling confirm the trade-off. Reducing anabolic drive can alleviate stress and slow decline, but at the cost of impaired repair and regeneration. Growth cannot be eliminated without compromising tissue integrity. Regulation must be partial and context-dependent.

Greed observed, therefore, is not excessive growth per se. It is growth operating without proportionate maintenance. Construction outpaces correction. Accumulation replaces renewal.

This imbalance does not immediately destroy tissues. It stabilises them in altered forms, denser, stiffer, less adaptable. The cost emerges later, in reduced flexibility and increased susceptibility to competitive takeover.

Greed does not act alone.

It amplifies the consequences of Sloth and prepares the ground for Pride.

Part II - Observations and Stress-Tests

II.3 Pride Observed

Persistence, senescence, and altered tissue ecology

The persistence of cells beyond their period of functional contribution is not an abstract concept. It is now directly measurable across ageing tissues.

Among the most widely documented manifestations of cellular persistence is senescence. Senescent cells are characterised by permanent withdrawal from the cell cycle coupled with resistance to apoptosis. They accumulate with age in multiple tissues, including skin, adipose tissue, lung, liver, and vasculature. Their abundance correlates with functional decline, fibrosis, and chronic inflammation, even in the absence of overt pathology.

Importantly, senescence arises as a protective response. Cells enter this state in reaction to irreparable damage, oncogenic stress, or replicative exhaustion. By halting proliferation, senescence prevents malignant expansion. In early life and acute injury, this response is adaptive and tightly regulated.

With age, its resolution falters.

Ageing tissues exhibit an increasing burden of senescent cells that persist long after their initial protective role has passed. Clearance mechanisms, primarily immune-mediated, become less efficient. Senescent cells resist apoptosis and accumulate, occupying physical and functional space within tissues. Their presence alters local signalling environments, often through sustained secretion of cytokines, proteases, and growth-modifying factors.

This persistence reshapes tissue ecology.

Neighbouring cells exposed to senescent secretory profiles exhibit altered differentiation, impaired regenerative capacity, and increased tolerance of damage. Stem cell niches are disrupted. Extracellular matrices are remodelled in ways that favour rigidity over adaptability. The tissue remains intact, but its internal dynamics change.

These effects are not confined to senescence alone. Other forms of cellular persistence contribute similarly. Long-lived immune cells accumulate with age, skewing repertoires towards memory and away from naïveté. Fibroblasts persist in activated states following injury, promoting fibrosis rather than resolution. Endothelial cells exhibit altered turnover, compromising vascular responsiveness. Across cell types, survival is increasingly favoured over replacement.

This shift reflects a broader change in exit thresholds.

Ageing does not abolish cell death pathways; it raises the cost of invoking them. Cells tolerate higher levels of damage before withdrawing. The decision to persist becomes easier than the decision to leave. What was once a tightly enforced boundary becomes negotiable.

Cancer again illustrates the extreme end of this continuum. Malignant cells evade senescence, resist apoptosis, and exploit survival pathways that are already softened in aged tissues. Yet the groundwork is laid well before transformation. Ageing tissues increasingly reward persistence, even when function declines.

Interventions targeting persistent cells expose the underlying tension.

Selective clearance of senescent cells in animal models improves tissue function and delays aspects of age-related decline. However, indiscriminate or sustained clearance impairs wound healing, disrupts tissue architecture, and compromises regenerative responses. Senescent cells are not uniformly harmful; their context matters. Removing persistence too aggressively destabilises tissues that rely on limited renewal.

From an ecological perspective, senescence and related persistence states alter the competitive landscape. Persistent cells occupy niches, consume resources, and shape signalling environments. They reduce turnover and increase heterogeneity. These conditions favour cells capable of enduring stress and resisting exit, traits that align with later clonal dominance.

Thus, Pride observed at the tissue level is not merely the accumulation of damaged cells. It is a systemic shift in how tissues negotiate survival versus replacement. Persistence becomes the default strategy. Renewal becomes conditional.

This shift does not immediately produce disease. In many cases, it stabilises ageing tissues sufficiently to maintain function. The cost is deferred. Reduced flexibility, impaired regeneration, and altered competitive dynamics accumulate quietly, increasing vulnerability to subsequent perturbation.

Pride, as observed, is therefore neither error nor excess. It is the predictable outcome of systems that must preserve structure while tolerating damage over extended periods.

Exit is not abandoned. It is postponed, often for too long.

Part II - Observations and Stress-Tests

II.4 When Ageing becomes cancer risk

Incidence, Time, and the limits of prevention

Cancer is often described as a disease of mutation. At the level of individual cells, this is correct. At the level of populations, it is incomplete.

Across tissues and tumour types, cancer incidence rises sharply with age. This rise is not linear. It accelerates after midlife, varies by tissue, and eventually plateaus or declines at extreme ages. These curves are remarkably consistent across populations, even as absolute risk shifts with environment, exposure, and detection practices.

The shape of these curves demands explanation.

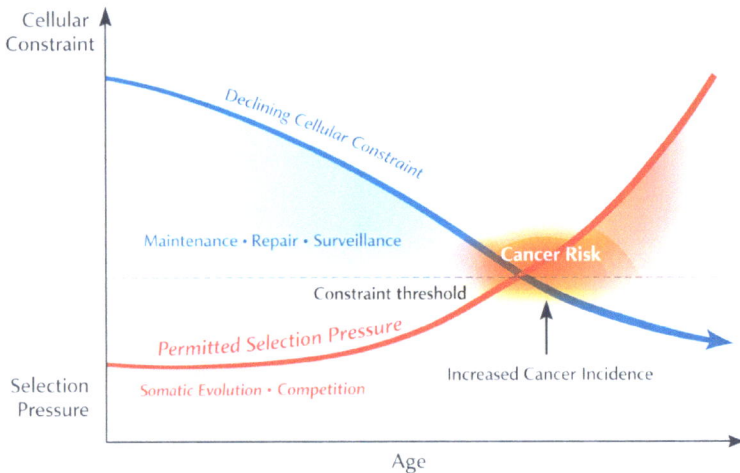

Ageing as a transition from cellular constraint to permitted selection. With age, mechanisms enforcing cellular cooperation gradually relax, allowing within-body selection to re-emerge. Cancer represents an extreme outcome of this process rather than its cause.

If mutation accumulation alone were sufficient, risk would be expected to increase more steadily with time, tracking cumulative exposure. Instead, incidence reflects a convergence of processes that unfold unevenly across the lifespan. Age is not merely a proxy for damage; it is a marker for changing tissue context.

Epidemiology captures this distinction indirectly. Cancer risk correlates more strongly with age than with any single exposure, even in tissues with high mutational burden early in life. Conversely, some tissues accumulate mutations extensively without corresponding cancer incidence. Mutation is necessary, but it is not decisive on its own.

What ageing alters is the environment in which mutations operate.

With advancing age, maintenance systems soften, growth control becomes permissive, clearance falters, and inflammatory tone rises. Metabolic signalling saturates. Replicative limits blur. Tissues become heterogeneous. In this altered landscape, mutations that would once have been neutral or deleterious acquire selective value. Risk accelerates not because mutations suddenly appear, but because they begin to matter.

This interpretation aligns with the timing of cancer emergence. Many oncogenic mutations are detectable years, sometimes decades, before diagnosis. Their presence alone does not predict disease. Progression depends on changes in tissue architecture, immune surveillance, and competitive dynamics, all of which shift with age.

Population-level data reinforce this view. Cancer incidence rises most steeply in tissues with high turnover and complex stem cell hierarchies, where opportunities for selection are greatest. Tissues with limited renewal show different patterns, despite comparable mutation rates. These differences are not explained by exposure alone. They reflect how ageing reshapes constraint.

Prevention strategies reveal the same logic.

Reducing exposure to known carcinogens lowers cancer risk, but does not eliminate age dependence. Enhancing early detection shifts diagnosis earlier, but does not flatten incidence curves. Even dramatic reductions in specific risk factors leave the overall age-related pattern largely intact. The slope changes; the shape remains.

This persistence suggests that ageing itself is the dominant risk modifier, not as a cause in isolation, but as the condition that allows other causes to act. Ageing transforms latent variation into realised disease.

The plateauing of cancer incidence at extreme ages further supports this interpretation. As tissues lose regenerative capacity and cellular competition intensifies, opportunities for malignant expansion diminish. Risk does not increase indefinitely because the environment eventually becomes hostile even to aggressive clones. Selection continues, but its outcomes change.

From an epidemiological perspective, cancer is therefore best understood not as an accumulation problem, but as a timing problem. Mutations accumulate throughout life. Cancer emerges when tissue context permits their exploitation. That permission is age-dependent.

This reframing clarifies why cancer and ageing are inseparable without being identical. Ageing does not programme cancer. It relaxes the constraints that suppress it. Prevention can delay this relaxation, but cannot abolish it without compromising other functions.

The hinge, then, is not mutation rate, but context.

Cancer risk rises when the balance between cooperation and competition tips inward, a transition that occurs gradually, unevenly, and predictably with age. Epidemiology does not merely record this transition. It reveals its inevitability.

II.5 Wrath Observed

Chronic Inflammation as failed resolution

Chronic inflammation is one of the most consistent correlates of ageing across tissues and species. Unlike acute inflammatory responses, which are rapid, localised, and self-limiting, age-associated inflammation is diffuse, persistent, and often clinically silent. Its presence is detectable not through overt symptoms, but through sustained elevations in inflammatory mediators and long-lived immune cell infiltration.

This pattern reflects a failure of resolution rather than excessive activation.

In youthful tissues, inflammatory responses are tightly coupled to repair. Damage triggers immune recruitment; debris is cleared; pro-resolving signals restore baseline. With age, this sequence becomes uncoupled. Inflammatory pathways are repeatedly engaged by low-level, unresolved stimuli, damaged macromolecules, persistent cells, altered extracellular matrices, without achieving closure.

The result is a state of chronic immune engagement without a clear endpoint.

Ageing tissues show increased infiltration of innate immune cells, altered macrophage phenotypes, and sustained cytokine production even in the absence of infection. Adaptive immune compartments also change. Memory populations expand, naïve cell pools contract, and immune responses become both exaggerated and ineffective. Surveillance persists, but precision declines.

This inflammatory milieu reshapes tissue behaviour.

Cells exposed to chronic inflammatory signals alter differentiation programmes, metabolic priorities, and stress responses. Stem cell niches become distorted. Regenerative cues weaken. Fibrotic pathways are favoured over restoration. Inflammatory mediators that are protective in the short term become disruptive when sustained.

Importantly, inflammation does not arise independently. It is reinforced by the sins that precede it.

Persistent senescent cells secrete pro-inflammatory factors that maintain immune activation. Maintenance failures increase the burden of debris requiring clearance. Metabolic saturation amplifies inflammatory signalling. Each contributes to a feedback loop in which inflammation becomes both consequence and cause of tissue dysfunction.

Cancer again occupies the extreme.

Tumours do not merely coexist with inflammation; they actively exploit it. Inflammatory environments promote angiogenesis, suppress effective immune clearance, and facilitate invasion. Yet these tumour-supportive effects do not require malignant intent. They arise from inflammatory circuits evolved to manage injury and infection, operating inappropriately over long timescales.

From a population perspective, chronic inflammation amplifies cancer risk by altering tissue context rather than by inducing mutation directly. Inflammatory signals increase cellular turnover, promote survival of damaged cells, and relax local constraints. Mutations that would otherwise remain inconsequential gain selective value. Risk rises not because inflammation creates cancer, but because it sustains conditions favourable to selection.

Interventions targeting inflammation reveal the same constraint.

Broad anti-inflammatory strategies can reduce tissue damage and alleviate symptoms, but they also impair host defence and tissue repair. Suppressing immune activity increases susceptibility to infection and malignancy. Enhancing immune activation risks autoimmunity and collateral damage. Resolution, not suppression, is the missing element, and resolution depends on restoring tissue conditions that ageing has compromised.

At the tissue level, chronic inflammation represents an ecological shift. Immune cells remain present, signals remain elevated, and the distinction between damage and baseline blurs. Defence becomes background noise. Regulation becomes probabilistic rather than precise.

Wrath observed, therefore, is not an overreaction. It is a response operating in an environment that no longer permits resolution. The immune system continues to act because the signals demanding action never fully disappear.

This persistence accelerates the transition from cooperation to competition. Inflammatory environments reward cells that tolerate stress, resist exit, and exploit altered signalling. Diversity narrows. Dominant clones gain ground.

Wrath does not initiate decline.

It magnifies it.

II.6 Gluttony Observed

Metabolic saturation and signal degradation

Metabolic excess in ageing tissues is not simply a matter of increased energy availability. It is a matter of signalling persistence.

Across organs, ageing is accompanied by sustained elevation of nutrient and growth-associated signals, glucose, lipids, amino acids, insulin, and related hormones, that were evolutionarily designed to fluctuate. These signals normally convey information about opportunity and demand. When they remain chronically elevated, their informational content degrades.

This degradation is observable at multiple levels.

At the cellular scale, prolonged exposure to excess nutrients induces adaptive resistance. Insulin and growth factor signalling pathways downregulate responsiveness to protect cells from overload. Initially, this preserves function. Over time, it blunts discrimination. Cells require stronger signals to achieve the same outcomes, increasing background activity while reducing precision.

At the tissue scale, this loss of sensitivity reshapes coordination. Metabolic cues that once synchronised growth, repair, and maintenance become decoupled from need. Anabolic processes persist even as quality control weakens. Clearance and repair lag behind synthesis. The result is accumulation rather than renewal.

Mitochondrial behaviour reflects this shift. In nutrient-rich environments, mitochondria adapt to sustained substrate availability by altering dynamics, efficiency, and redox balance. These adaptations maintain ATP production but increase metabolic by-products and stress signalling. Energy continues to flow, but feedback becomes noisy. The cell remains active without being well-regulated.

Importantly, metabolic saturation does not act in isolation. It interacts with the preceding sins.

Persistent cells permitted by Pride exploit abundant resources. Chronic inflammation associated with Wrath amplifies metabolic signalling and alters substrate use. Maintenance failures associated with Sloth increase susceptibility to metabolic stress. Together, these conditions create environments in which excess is tolerated but poorly managed.

From an ecological perspective, metabolic saturation alters selective pressures.

Cells capable of tolerating signal noise, exploiting alternative substrates, or uncoupling growth from regulation gain advantage. Precision becomes less valuable than endurance. Under these conditions, cells that maintain tight coupling between signal and function are disadvantaged. Selection favours robustness over accuracy.

Cancer again illustrates the extreme.

Tumour cells thrive in metabolically saturated environments by narrowing interpretive range rather than restoring sensitivity. They upregulate transporters, reroute central carbon metabolism, and exploit lipids and amino acids when glucose handling falters. Where normal tissues experience confusion, malignant clones impose coherence, on their own terms.

Yet the underlying phenomenon precedes malignancy. Ageing tissues without cancer show altered substrate utilisation, ectopic lipid accumulation, and persistent activation of growth-associated pathways despite declining functional output. Metabolism continues, but its coordination with repair and replacement weakens.

Interventions that reduce metabolic load demonstrate partial reversibility. Improving insulin sensitivity or reducing nutrient excess can restore some signalling fidelity and tissue responsiveness. However, these effects are often transient and context-dependent. Systems adapted to saturation recalibrate rather than reset. Signal clarity improves only within limits.

This constraint reflects a broader principle. Metabolism functions as an information-processing system as much as an energy-delivery system. Its signals acquire meaning only through contrast, through fluctuation between abundance and scarcity. When abundance becomes continuous, signalling saturates and loses specificity.

Gluttony observed, therefore, is not overconsumption alone. It is the persistence of abundance beyond interpretive capacity. Signals remain present, but they no longer guide behaviour reliably.

In such environments, regulation yields to persistence, coordination yields to tolerance, and selection favours cells that can function amid noise. These conditions accelerate the transition from cooperative tissue behaviour to competitive dynamics.

Gluttony does not initiate decline.

It ensures that, once begun, decline is metabolically sustained.

Part II - Observations and Stress-Tests

II.7 Lust Observed

Replicative persistence and stem-cell drift

Replicative persistence in ageing tissues is neither uniform nor absolute. It is concentrated within specific compartments, most notably stem and progenitor cell pools, where continued division is essential for tissue maintenance. What changes with age is not the existence of replication, but its distribution, regulation, and consequence. Where Pride concerns cells that persist without dividing, Lust concerns cells that persist by continuing to divide.

In youthful organisms, stem cell compartments balance self-renewal with differentiation. Replication preserves tissue integrity while limiting the duration of clonal persistence. Lineages turn over. Diversity is maintained. Replicative potential is allocated, not accumulated.

With age, this balance shifts.

Multiple tissues exhibit evidence of stem-cell drift: the gradual dominance of a subset of long-lived clones within renewal compartments. These clones are not defined by overt transformation, but by relative advantages in survival, quiescence, or stress tolerance. Over time, they contribute disproportionately to tissue maintenance, while alternative lineages diminish or disappear.

This phenomenon is now documented across organ systems.

In haematopoietic tissues, ageing is accompanied by reduced clonal diversity and expansion of long-lived stem-cell lineages with altered differentiation bias. In the intestine, crypts become increasingly monoclonal with age, reflecting clonal replacement rather than equilibrium. In skin, hair follicles, and liver, similar patterns emerge: renewal continues, but from a narrowing pool of contributors.

Replicative persistence thus becomes uneven.

Cells capable of maintaining replicative capacity under stress, through enhanced DNA damage tolerance, altered checkpoint control, or resistance to senescence, gain advantage. These traits do not initially compromise tissue function. In many cases, they stabilise it. The cost is paid in reduced adaptability and increased vulnerability to subsequent perturbation.

Telomere dynamics reflect this trade-off. In dividing compartments, telomere shortening constrains replication, enforcing turnover. With age, cells that preserve telomere length or tolerate telomere dysfunction persist longer. This persistence supports continued renewal but also permits the retention of accumulated damage. Replicative limits soften without disappearing.

Cancer again represents the extreme outcome of this process.

Most tumours secure replicative persistence by activating telomere maintenance mechanisms or by tolerating genomic instability. Yet these strategies exploit pathways already present in normal renewal compartments. The difference lies not in mechanism, but in allocation. What is tightly restricted in healthy tissues becomes broadly available under malignant selection.

Importantly, replicative persistence does not require increased division rates. In many ageing tissues, overall proliferation declines. What increases is lineage duration. Cells divide fewer times per unit time, but remain contributors for longer periods. Drift replaces turnover. Persistence replaces succession.

Interventions targeting replicative persistence illustrate the constraint.

Enhancing stem-cell self-renewal can delay exhaustion and preserve tissue function, but increases clonal dominance and instability. Enforcing stricter replicative limits reduces cancer risk, but accelerates tissue failure. Replicative persistence is therefore not a defect to be eliminated, but a compromise to be managed.

From an ecological perspective, stem-cell drift narrows the competitive field. Diversity declines. The tissue becomes dependent on fewer lineages. Under these conditions, additional advantages, metabolic tolerance, resistance to inflammation, evasion of clearance, carry greater weight. Selection intensifies.

Lust observed, then, is not the pursuit of immortality by all cells. It is the uneven extension of contribution by a few. Replication persists where replacement falters. Continuity is maintained at the expense of diversity.

This persistence stabilises ageing tissues temporarily.

It also prepares them for dominance.

II.8 Envy Observed

Clonal Selection in ageing tissues

The return of competition within ageing tissues is no longer a theoretical concern. It is now directly observable.

Over the past decade, large-scale sequencing studies have revealed that many tissues in older individuals are not cellular mosaics of near-equal participants, but patchworks dominated by expanding clones. These clones are defined not by overt malignancy, but by the possession of modest advantages in survival, proliferation, or stress tolerance. Their presence reflects selection acting within tissues whose regulatory constraints have softened with age.

The most thoroughly characterised example is the ageing blood system. Deep sequencing of peripheral blood from large human cohorts has shown that a substantial fraction of individuals over the age of sixty harbour expanded haematopoietic clones bearing mutations in genes associated with growth regulation, epigenetic control, or DNA damage response. These clones often comprise a significant proportion of circulating blood cells, yet produce no immediate clinical abnormality. They are not cancer. They are selection made visible.

What distinguishes these clones is not aggression, but persistence. Mutations that subtly enhance self-renewal, resist apoptosis, or improve tolerance to inflammatory environments allow affected stem or progenitor cells to outcompete their neighbours. Over time, their descendants displace other lineages. Diversity contracts. The tissue remains functional, but its composition shifts.

Similar patterns are now evident in solid tissues.

In the skin, sequencing of ostensibly normal epithelium reveals extensive clonal expansions carrying mutations in genes that regulate proliferation and differentiation. These clones can span millimetres of tissue without breaching histological definitions of normality. In the oesophagus, ageing is accompanied by near-complete replacement of the epithelial lining by a small number of dominant clones, many bearing mutations identical to those seen in squamous cell carcinoma. Most never progress to cancer. All reshape the tissue environment.

These observations confirm a central prediction of the sin framework: Envy does not require malignancy. It requires only variation and opportunity.

Ageing supplies both. Accumulated damage, chronic inflammation, metabolic stress, and impaired clearance create heterogeneous microenvironments in which small differences between cells are amplified. Regulatory systems that once suppressed selection, uniform growth control, efficient exit, immune surveillance, become less precise. Under these conditions, even marginal advantages are sufficient to drive clonal dominance.

Importantly, this process is not uniformly deleterious in the short term. Dominant clones often preserve tissue integrity more effectively than their failing neighbours. Selection initially stabilises function. The cost is paid later, in reduced diversity and increased vulnerability to further perturbation.

This duality complicates intervention.

Enhancing surveillance or clearance can limit clonal expansion, but at the cost of increased selective pressure. Cells that evade detection or tolerate stress are preferentially retained. Selection accelerates. Conversely, relaxing control to preserve diversity permits the expansion of the most persistent clones. Either approach alters tempo, not direction.

Cancer prevention strategies encounter this constraint repeatedly. Reducing mutation burden, enhancing DNA repair, or increasing turnover can delay the emergence of dominant clones, but cannot eliminate selection itself. As long as cells differ, and difference is unavoidable in living tissues, competition resumes when constraints weaken.

What ageing reveals, then, is not the sudden appearance of evolutionary dynamics, but their gradual re-entry into spaces from which they were once excluded. Multicellular order suppresses selection locally and temporarily. Ageing marks the erosion of that suppression.

The significance of these findings extends beyond cancer risk. Clonal dominance alters tissue behaviour even in the absence of transformation. Dominant clones shape local signalling, modify extracellular matrices, and influence immune interactions. They change not only who occupies the tissue, but how that tissue responds to stress, injury, and repair.

Envy, in its observed form, is therefore not a late catastrophe but an early drift. Cancer is merely the most visible outcome of a process that begins decades earlier, operating quietly within the boundaries of apparent normality.

This observation closes the empirical loop.

What Part I described as moral metaphor, competition reasserting itself within cooperation, is now measurable, quantifiable, and widespread. The truce does not fail suddenly. It erodes, cell by cell, clone by clone, under conditions shaped by time.

Selection does not return because control disappears.

It returns because control was never free.

Part II - Observations and Stress-Tests

II.9 Why Optimisation fails

Trade-offs, constraints, and the structure of limits

Across the preceding observations, a consistent pattern emerges. Interventions that improve one dimension of tissue function reliably impose cost elsewhere. Gains are real, but they do not accumulate. Instead, they redistribute burden across time, cell types, or functions.

This is not a failure of implementation.

It is a consequence of structure.

Biological systems are not optimised for maximal performance along any single axis. They are shaped to remain viable under uncertainty. Maintenance, growth, clearance, defence, metabolism, and renewal are balanced against one another under constraints imposed by energy, time, and evolutionary history. Improving one component necessarily alters the equilibrium of the others.

This is why optimisation fails as a strategy.

Attempts to enhance maintenance reduce damage burden but increase energetic cost and reduce flexibility. Suppressing growth alleviates stress but compromises repair and regeneration. Enforcing exit clears dysfunctional cells but destabilises tissue architecture. Dampening inflammation reduces collateral damage but weakens defence. Sharpening metabolic signalling improves precision but reduces tolerance to stress. Limiting replicative persistence constrains malignancy but accelerates exhaustion.

Each intervention succeeds locally.

Each fails globally.

These failures are not symmetric. The costs are often delayed, diffuse, or context-dependent. Early benefits can obscure later consequences. This temporal displacement creates the illusion that optimisation is working, until accumulated trade-offs surface elsewhere in the system.

Ageing makes these trade-offs visible.

In youthful organisms, reserves of redundancy and adaptability buffer imbalance. Costs can be absorbed without overt consequence. With age, these buffers thin. Interventions that once improved function now expose fragility. The margin for correction narrows. Optimisation becomes destabilising.

Cancer prevention illustrates this clearly. Measures that reduce cancer risk, limiting proliferation, enhancing surveillance, enforcing exit, also impair renewal and resilience. Conversely, preserving regenerative capacity increases long-term risk. There is no intervention that improves both indefinitely. The apparent conflict reflects the underlying architecture of multicellular life.

At a deeper level, optimisation fails because it presumes independence where none exists.

Cells are not modular units. Tissues are not collections of interchangeable parts. Functions overlap, share resources, and constrain one another. Altering one parameter reshapes the landscape in which all others operate. Biology responds not by complying, but by rebalancing.

This rebalancing often takes the form of selection.

When constraints tighten unevenly, cells that tolerate altered conditions persist. Clonal dominance increases. Diversity contracts. Stability is regained locally at the cost of global adaptability. Optimisation accelerates the very dynamics it seeks to suppress.

This is why ageing cannot be reduced to a single target, and why cancer cannot be prevented by eliminating one pathway. The system adapts around pressure. Control provokes compensation. Improvement invites rearrangement.

None of this implies that intervention is futile.

Rates can be slowed. Burdens can be shifted. Periods of function can be extended. Suffering can be reduced. But these outcomes are achieved by *managing trade-offs*, not by abolishing them. Optimisation fails because biology is not built to be perfected.

It is built to persist under compromise.

This observation prepares the ground for the final perspective.

If limits are structural rather than accidental, then the task is not to eliminate them, but to understand how they shape what is possible, and what is not. Only from that understanding can restraint, rather than optimisation, become a coherent strategy.

Part II closes here.

What follows is not evidence, but perspective.

Trade-Offs Prevent Global Optimisation

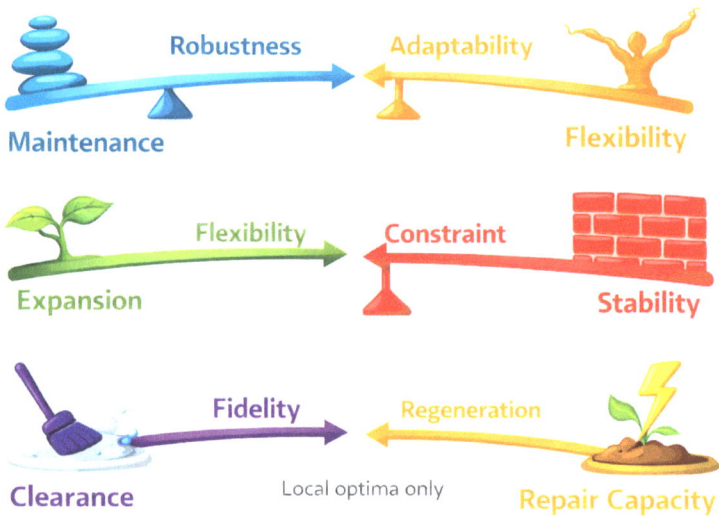

Structural trade-offs prevent global optimisation in biological systems. Enhancing maintenance, expansion, or clearance inevitably redistributes cost to flexibility, constraint, or repair capacity. Interventions therefore rebalance system priorities rather than eliminate underlying limits, enforcing local rather than global optima.

Part III

Redemption Without Illusion

I. Redemption as Rebalancing, Not Reversal

In moral traditions, redemption implies return, a restoration of what was lost. Biology permits no such return.

Cells retain history. Tissues bear accumulation. Selection, once reintroduced, does not retreat. Ageing cannot be undone in the sense that time can be reversed or error erased. Any redemption premised on reversal mistakes the nature of the system.

What biology does permit is rebalancing.

Throughout this work, decline has been described as the relaxation of constraints that sustain cooperation. That relaxation is uneven. Some processes soften earlier than others. Some tissues lose restraint while others retain it. This unevenness creates vulnerability, but it also creates leverage, not for perfection, but for proportion.

Redemption, in biological terms, does not mean abolishing the sins. It means preventing their convergence.

Sloth is redeemed not by exhaustive repair, but by restoring sufficient maintenance to prevent runaway heterogeneity.
Greed is redeemed not by suppressing growth, but by re-aligning synthesis with repair.
Pride is redeemed not by enforcing universal exit, but by re-enabling selective clearance.
Wrath is redeemed not by silencing immunity, but by restoring resolution.
Gluttony is redeemed not by deprivation, but by reintroducing signal contrast.
Lust is redeemed not by forbidding persistence, but by redistributing contribution.

Envy is redeemed not by eliminating competition, but by delaying its escalation.

Each act is partial. Each carries cost. None offers absolution.

Together, they describe a *modus vivendi*, a way of living negotiated under constraint.

II. The Biological Meaning of Discipline

Discipline, as biology understands it, is not austerity. It is **selectivity**.

Living systems function best when pressures are applied locally, transiently, and incompletely. Acute stress strengthens; chronic stress degrades. Absolute suppression destabilises; unchecked permissiveness dissolves coherence. The same principle recurs at every scale examined in this book.

This gives redemption a precise meaning.

Interventions that succeed are those that:

- act early rather than late
- restore dynamic range rather than fixed states
- preserve heterogeneity rather than enforce uniformity
- accept partial benefit without total control

These are not ideals imposed from outside biology. They are properties of resilient systems: pulsed repair, transient growth, selective clearance, resolved inflammation, fluctuating metabolism, bounded renewal. Youthful tissues exhibit them naturally. Ageing tissues lose them gradually.

The optimism here is real, but narrow.

It lies not in escaping the architecture of ageing, but in operating within it with restraint. Discipline resists the temptation to drive systems to extremes in the name of progress. It prefers coherence to maximal performance, adaptability to endurance, and stability to optimisation.

The principle is ancient: *ne quid nimis*.

Nothing in excess.

III. Redemption at the Level of Action

Once understood, the seven sins cease to be merely descriptive. They become diagnostic.

They allow us to ask, of any intervention:
Which sin does this restrain?
Which does it amplify?
Which costs are advanced, and which deferred?

This reframes progress. Success is no longer measured by elimination of decline, but by shaping its trajectory. The aim is extension of function without acceleration of collapse, maintaining recognisable order for as long as possible under constraint.

Such redemption is modest in promise and exacting in practice. It offers no decisive breakthrough, no final victory. It offers something rarer: **the ability to act without self-deception**.

This is an optimistic position, but not a consoling one.

It accepts that every gain carries shadow, that every correction reshapes the field, and that the absence of catastrophe is itself an achievement. Knowledge, applied with restraint, can slow disintegration even when it cannot abolish it.

IV. After the Sins

The seven cellular sins are not moral failures. They are the predictable consequences of life sustained over time.

Redemption, therefore, does not consist in erasing them, but in **preventing their alignment**. Ageing becomes destructive when sloth enables greed, greed sustains pride, pride fuels wrath, wrath amplifies gluttony, gluttony supports lust, and lust accelerates envy. Redemption lies in breaking these cascades, repeatedly, imperfectly, and consciously.

This is the deepest optimism biology allows.

Not that decline can be escaped, but that it can be **inhabited without surrendering coherence**. Not that ageing can be defeated, but that its most destructive accelerations can be resisted. Not that life can be made permanent, but that it can remain recognisably itself for longer.

After the sins, there is no absolution.

There is *modus vivendi*.

That is discipline.

Time does not undo living systems by catastrophe, but by account. What is borrowed early is repaid later, often quietly, and often elsewhere. Biology keeps its books without malice or exception.

Ageing is not the failure of vigilance, nor the triumph of disorder. It is the price paid for cooperation maintained under constraint. That price is not collected all at once, but gradually, as precision gives way to tolerance and restraint thins without vanishing.

In this sense, time does not err. *Tempus non errat.* What unravels does so according to rules that were present from the start.

Appendix

From Selection to Suppression

A Short History of Multicellularity, Ageing, and Cancer

Modern biology inherited its central insight from the nineteenth century: that living systems are shaped by selection acting on variation over time. When **Charles Darwin** articulated this principle, he was careful to define its scope. Natural selection operates on immediate advantage. It rewards traits that improve survival or reproduction in the present environment. It has no foresight, and no mechanism for privileging distant outcomes over proximal success.

This limitation is not a weakness of the theory. It is its defining feature.

The consequences of this logic become most visible when selection is displaced from its usual level of action. Multicellular life represents such a displacement. For the first time, selection acting on individual cells had to be restrained so that selection acting on organisms could prevail. Cells capable of unchecked proliferation, exploitation of resources, or resistance to death were no longer advantageous. They became threats to the collective.

Multicellularity therefore required something unprecedented: the **suppression of selection within the body**.

This suppression was never total. It was achieved through layered constraints, developmental programmes, lineage commitment, growth control, programmed cell death, and immune surveillance. These mechanisms did not eliminate variation or competition; they managed it. They allowed organisms to develop, maintain form, and reproduce by constraining the autonomy of their cellular constituents.

From an evolutionary perspective, this was a precarious arrangement. The same traits that enabled multicellular success, plasticity, repair, regeneration, also preserved the potential for cellular divergence. Selection was deferred, not abolished.

Ageing emerges from this deferral.

Because selection prioritises early-life success, maintenance systems evolved to be sufficient rather than exhaustive. Repair pathways correct common damage efficiently, but rare, cumulative, or late-arising failures are tolerated. Surveillance is effective when it matters most for reproductive success, and less so thereafter. Restraint is enforced while its benefits outweigh its costs.

The result is not programmed decay, but **declining precision**. Over time, constraints soften unevenly. Repair slows. Clearance falters. Signals persist beyond their optimal range. Cells that would once have been eliminated remain. None of this requires failure in the everyday sense. It follows directly from how selection allocates investment.

Cancer represents the most explicit consequence of this allocation.

Historically, cancer was framed as a disease of excess proliferation, then as a genetic disease, and later as a disorder of signalling. Each framing captured part of the phenomenon. What they shared was an implicit assumption that cancer was an aberration, an intrusion into otherwise stable biology.

A different picture emerges when cancer is viewed through the lens of evolutionary history.

Cancer is not the negation of multicellularity. It is its shadow.

The same cellular capacities that enable development and regeneration, division, survival, adaptation, become liabilities when restraint weakens. Tumours do not invent new behaviours; they exploit existing ones under altered constraints. They are not foreign entities, but **rearrangements of priority** within the same system.

This perspective has been developed in detail elsewhere. In *The Hallmarks of Cancer: The New Testament*, the argument was advanced that cancer progression reflects not merely the acquisition of enabling traits, but a systematic erosion of the cooperative framework that defines multicellular order. The emphasis there was on expansion, how cancer capabilities accumulate and interact.

The present work approaches the same problem from the opposite direction.

Here, the focus is on **loss rather than gain**: the gradual loosening of restraint that precedes overt pathology. Where the hallmarks framework catalogues the capacities cancer cells acquire, the cellular sins described in this book identify the conditions that allow those capacities to become advantageous in the first place.

Seen together, these perspectives are complementary.

The hallmarks describe what cancer becomes.
The sins describe what ageing permits.

Both arise from the same evolutionary bargain.

Historical attempts to separate ageing and cancer as distinct biological problems have struggled for this reason. Interventions that suppress proliferation increase degenerative decline.

Strategies that preserve regeneration elevate cancer risk. Enhancing surveillance improves early-life robustness at the expense of later-life fragility. These trade-offs are not technical failures; they are structural consequences of how multicellular life is organised.

What has changed in recent decades is not the logic, but our ability to observe it directly.

Clonal expansions in normal tissues, age-associated shifts in immune function, metabolic saturation, senescent cell accumulation, and altered stem-cell dynamics now provide empirical confirmation of processes that evolutionary theory long implied. Ageing is no longer inferred solely from organismal decline. It is visible at the level of cell populations and tissue ecology.

This convergence of theory and observation has sharpened the central question.

If ageing and cancer both emerge from the same compromise, then neither can be fully understood, or effectively addressed, in isolation. Efforts to eliminate cancer without regard to ageing, or to extend lifespan without regard to cancer risk, will inevitably encounter the same limits from opposite directions.

The historical lesson is therefore not pessimistic, but clarifying.

Multicellular life is possible because selection within the body is restrained.
Ageing occurs because that restraint is costly to maintain indefinitely.
Cancer arises when restraint weakens sufficiently for competition to reassert itself.

These are not separate stories. They are stages of one.

This appendix does not add a new argument to the book. It situates the argument within a longer intellectual lineage, from Darwin's original insight into selection, through the evolution of multicellular constraint, to contemporary understanding of cancer as an evolutionary process within the body.

The seven cellular sins described in the main text are not deviations from this history. They are its modern expression.

Cancer as Exploitation, not Invention

A recurring misconception in the history of oncology has been the assumption that cancer invents new biological behaviours. On the contrary, tumour cells rarely create novel mechanisms. They repurpose existing ones.

This principle has been articulated most clearly in the modern refinement of the hallmarks framework. In *Hallmarks of Cancer: The New Testament*, cancer is described not as a disease defined by singular mutations, but as a progressive reconfiguration of normal cellular capabilities, plasticity, epigenetic modulation, microbial interaction, and neuronal signalling, deployed outside their proper constraints (Senga & Grose 2021).

What is crucial in that account is not the enumeration of traits, but the underlying logic: tumour cells succeed precisely because they exploit programmes that are already present, already functional, and already adaptive under other circumstances. Dedifferentiation recapitulates developmental plasticity. Epigenetic dysregulation leverages mechanisms evolved to permit context-dependent gene expression. Microbial and neuronal interactions reflect long-standing co-evolutionary relationships rather than pathological novelty.

The present work approaches the same biological reality from a complementary direction.

Where the hallmarks framework catalogues what cancer cells acquire, the cellular sins described in this book identify what ageing permits. They describe the **progressive relaxation of the constraints** that normally keep those capabilities proportionate, localised, and reversible.

Seen in this light, cancer is neither aberrant nor accidental. It is the predictable outcome of cellular autonomy reasserting itself as multicellular restraint weakens. Ageing supplies the permissive substrate; malignancy exploits it.

This distinction matters. It clarifies why efforts to target individual hallmarks often yield temporary control rather than durable resolution. The problem is not the presence of plasticity, growth, survival, or signalling. It is their persistence beyond context, their escape from collective discipline.

The hallmarks and the sins are therefore not competing frameworks. They describe different phases of the same evolutionary bargain. One enumerates the capabilities that emerge when constraint fails. The other explains why that failure is inevitable over time.

Reference:

Senga, S. S., & Grose, R. P. (2021). *Hallmarks of cancer-the new testament*. Open Biology, 11(1)**,** 200358.

www.ingramcontent.com/pod-product-compliance
Lightning Source LLC
Chambersburg PA
CBHW041712200326
41518CB00005B/193